Quantitative Thin-Layer Chromatography and Its Industrial Applications

CHROMATOGRAPHIC SCIENCE

A Series of Monographs

Editor: JACK CAZES
Silver Spring, Maryland

Volume 1: Dynamics of Chromatography
J. Calvin Giddings

Volume 2: Gas Chromatographic Analysis of Drugs and Pesticides
Benjamin J. Gudzinowicz

Volume 3: Principles of Adsorption Chromatography: The Separation of Nonionic Organic Compounds (out of print)
Lloyd R. Snyder

Volume 4: Multicomponent Chromatography: Theory of Interference (out of print)
Friedrich Helfferich and Gerhard Klein

Volume 5: Quantitative Analysis by Gas Chromatography
Joseph Novák

Volume 6: High-Speed Liquid Chromatography
Peter M. Rajcsanyi and Elisabeth Rajcsanyi

Volume 7: Fundamentals of Integrated GC-MS (in three parts)
Benjamin J. Gudzinowicz, Michael J. Gudzinowicz, and Horace F. Martin

Volume 8: Liquid Chromatography of Polymers and Related Materials
Jack Cazes

Volume 9: GLC and HPLC Determination of Therapeutic Agents (in three parts)
Part 1 edited by Kiyoshi Tsuji and Walter Morozowich
Part 2 and 3 edited by Kiyoshi Tsuji

Volume 10: Biological/Biomedical Applications of Liquid Chromatography
Edited by Gerald L. Hawk

Volume 11: Chromatography in Petroleum Analysis
Edited by Klaus H. Altgelt and T. H. Gouw

Volume 12: Biological/Biomedical Applications of Liquid Chromatography II
Edited by Gerald L. Hawk

Volume 13: Liquid Chromatography of Polymers and Related Materials II
Edited by Jack Cazes and Xavier Delamare

Volume 14: Introduction to Analytical Gas Chromatography: History, Principles, and Practice
John A. Perry

Volume 15: Applications of Glass Capillary Gas Chromatography
Edited by Walter G. Jennings

Volume 16: Steroid Analysis by HPLC: Recent Applications
Edited by Marie P. Kautsky

Volume 17: Thin-Layer Chromatography: Techniques and Applications
Bernard Fried and Joseph Sherma

Volume 18: Biological/Biomedical Applications of Liquid Chromatography III
Edited by Gerald L. Hawk

Volume 19: Liquid Chromatography of Polymers and Related Materials III
Edited by Jack Cazes

Volume 20: Biological/Biomedical Applications of Liquid Chromatography IV
Edited by Gerald L. Hawk

Volume 21: Chromatographic Separation and Extraction with Foamed Plastics and Rubbers
G. J. Moody and J. D. R. Thomas

Volume 22: Analytical Pyrolysis: A Comprehensive Guide
William J. Irwin

Volume 23: Liquid Chromatography Detectors
Edited by Thomas M. Vickrey

Volume 24: High-Performance Liquid Chromatography in Forensic Chemistry
Edited by Ira S. Lurie and John D. Wittwer, Jr.

Volume 25: Steric Exclusion Liquid Chromatography of Polymers
Edited by Josef Janča

Volume 26: HPLC Analysis of Biological Compounds: A Laboratory Guide
William S. Hancock and James T. Sparrow

Volume 27: Affinity Chromatography: Template Chromatography of Nucleic Acids and Proteins
Herbert Schott

Volume 28: HPLC in Nucleic Acid Research: Methods and Applications
Edited by Phyllis R. Brown

Volume 29: Pyrolysis and GC in Polymer Analysis
Edited by S. A. Liebman and E. J. Levy

Volume 30: Modern Chromatographic Analysis of the Vitamins
Edited by Andre P. De Leenheer, Willy E. Lambert, and Marcel G. M. De Ruyter

Volume 31: Ion-Pair Chromatography
Edited by Milton T. W. Hearn

Volume 32: Therapeutic Drug Monitoring and Toxicology by Liquid Chromatography
Edited by Steven H. Y. Wong

Volume 33: Affinity Chromotography: Practical and Theoretical Aspects
Peter Mohr and Klaus Pommerening

Volume 34: Reaction Detection in Liquid Chromatography
Edited by Ira S. Krull

Volume 35: Thin-Layer Chromatography: Techniques and Applications, Second Edition, Revised and Expanded
Bernard Fried and Joseph Sherma

Volume 36: Quantitative Thin-Layer Chromatography and Its Industrial Applications
Edited by Laszlo R. Treiber

Volume 37: Ion Chromatography
Edited by James G. Tarter

Other Volumes in Preparation

Quantitative Thin-Layer Chromatography and Its Industrial Applications

edited by

Laszlo R. Treiber
Chemical Engineering Research and Development Department
Merck & Co., Inc.
Rahway, New Jersey

Marcel Dekker, Inc. **New York and Basel**

Library of Congress Cataloging in Publication Data

Quantitative thin-layer chromatography and its industrial applications.

(Chromatographic science ; v. 36)
Includes index
1. Thin layer chromatography. 2. Chemistry, Analytic—Quantitative. I. Treiber, Laszlo R.
II. Series.
QD117.C52Q36 1987 544'.956 86-24231
ISBN 0-8247-7597-X

MARCEL DEKKER, INC.
270 Madison Avenue, New York, New York 10016

Current printing (last digit):
10 9 8 7 6 5 4 3 2 1

PRINTED IN THE UNITED STATES OF AMERICA

Preface

The first four chapters of this book discuss the general background of quantitative thin-layer chromatography (QTLC): materials used for separation of components and theoretical aspects of quantitation. Chapter 3 is devoted specifically to the applications in screening programs, for which some unique features of thin-layer chromatography (TLC) can best be demonstrated. Following are the special topics representing major industrial applications, each of which could fill a book on its own. The reader will find the concise, readily understandable style and the extensive literature references combined with the experience of the contributors' laboratories extremely valuable. All the authors are distinguished scientists recognized in the field of separation techniques, in general, and in the topic they present, in particular. It was a genuine pleasure working with the contributors, who made this book a well-balanced mixture of the objective facts of peaks and numbers and the somewhat subjective art of intuition and personal touch.

Whereas facts expressed as numerical data can be dealt with by computers or human intellect, intuition is exclusively human. Whenever these two factors merge in the right proportion, we have the most powerful combination for the advancement of science as well as for industrial applications. One of our aims was to demonstrate how QTLC fits in a well-managed analytical laboratory as a technique on its own but also as one complementary to the other chromatographic methods.

Finally, I would like to pay tribute to Professor G. W. Oertel, who some two decades ago taught me everything there was to know about chromatography at that time. As his laboratory research assistant, first in Homburg/Saar, then in Mainz, Federal Republic of Germany, I had the honor and unique opportunity to learn how to blend objective facts and subjective intuitions into something that could perhaps be termed the *process of thinking.* QTLC is certainly a manifestation of the scientific approach, currently richer in the subjective elements than any other chromatographic technique. No doubt my interest in QTLC originates from the years spent with Professor Oertel. It is rewarding to see that many ideas that started taking shape at that time in his and several other laboratories were subsequently converted to generally accepted practice.

Laszlo R. Treiber

Introduction

Before any method can be considered for industrial application, three basic requirements must be met. First, a technical concept that is compatible with the properties of the sample to be analyzed must exist. Second, the method must meet specified standards of precision, sensitivity, and specificity. Third, operating costs (equipment, parts, materials and supplies, and personnel expenses) must remain within reasonable limits.

As far as the first requirement is concerned, both quantitative liquid-column chromatography (QLC) and quantitative thin-layer chromatography (QTLC) fall essentially within the same areas. Both concepts belong to the same technical category: liquid chromatography in a broader sense. The physical phenomena by which differential migration is achieved are virtually identical in both cases as are the scope and limitations. However, differences in equipment design, such as mechanical components supporting the stationary phase and facilitating the flow of the mobile phase, may have a decisive influence on the second and third requirements. In meeting the second requirement, the criteria of reliability, QLC is generally superior over QTLC. For the same amount of sample, liquid chromatography (LC) detection is usually more accurate, precise, and sensitive than QTLC. The specificity, however, is probably equal or even more favorable in QTLC. Comparisons on the basis of the above criteria might give different results for different cases and reflect only the currently existing status of technical development. Future advancements in instrument design are likely to favor QTLC more than QLC, since the technical and financial efforts invested into the latter have brought its capabilities very close to the theoretical limits. But even in its present state of technical development, QTLC satisfies the needs of many applications.

The third requirement to be met, reasonable operating costs, has bearing not only on technical aspects but also on the manufacturing costs of a product requiring analytical work. The manager of an industrial laboratory has to work within the budget allocated to analytical services. The appropriate balance of instrumentation is essential in terms of capital investment. The flexibility of QTLC is potentially an important factor in cutting funds spent for instrumentation. Automated QTLC systems have a sample throughput unparalleled by any other chromatographic technique, and we are not even close to the capacity attainable on the basis of the current state of available technology.

Certainly, there are limits in the performance of QTLC just as in any other technique. We must know those limits in order to make reliable evaluations of experimental data and especially to optimize the application of QTLC in laboratory practice.

Many of the characteristics unique to QTLC are discussed by the authors of the individual chapters. To avoid redundancy, it is appropriate to collect their common views for presentation in this introduction.

The most important and a unique feature of QTLC is the fact that it operates on a *time sharing* basis. Unlike instruments for liquid or gas chromatography, the QTLC instrument is not dedicated to any individual sample, or even to any specific assay procedure during the development cycle of the chromatography. Any number of samples can be separated simultaneously, independently in space and time, by means of any number of thin-layer chromatographic methods. The rate of scanning can typically be less than a minute per sample. However, the technology is available for scanning a sample in only a few seconds. The rate-limiting factor for sample throughput is the speed of quantitative scan and not the speed of chromatography.

Another unique advantage of QTLC is the freedom to select any developing solvent volatile enough to evaporate after chromatography. In contrast to QTLC, serious restrictions exist in QLC: only solvents sufficiently transparent at the wavelength of detection can be used.

Furthermore, in QTLC every sample is separated on fresh adsorbent, hence carryover and cross-contamination are avoided. Unlike in LC, fractions irreversibly adsorbed can still be detected. The options for sample preparation are generally less demanding and less expensive. Thin-layer chromatography plates can accept suspended solid particles, aggressive agents such as acids or bases in the sample, that would lead to the certain destruction of many types of LC columns and other instrument components. This ruggedness of TLC plates allows savings in operating expenses, time, and material.

Postchromatography derivatization is inexpensive and much more versatile in TLC than in LC, regardless of its purpose. This technique in LC usually requires additional instrument components potentially as expensive as the basic LC instrument itself. In TLC simple and inexpensive spraying devices, dipping tanks, and chambers facilitating reaction with gaseous reagents serve the purpose of postchromatography derivatization. Quantitative methods based on spray and dipping reagents are generally less reliable than postcolumn derivatization methods in LC. However, the choice of reagents is far greater in TLC than in LC. This translates to higher specificity in QTLC.

Here I would like to pause for a few personal remarks. During a scientific meeting a fellow chromatographer, whose name I do not remember, stated that he liked QTLC much more than QLC because he could *see* the chromatogram. He pointed out a very important feature of QTLC: the *visual* experience. Some 90% of all the information we gather reaches us through our visual sense. The characteristics of a TLC spot, such as its color and shape or the color reactions with different spray reagents at various temperatures or when exposed to the air or other gases, are usually unique for a compound. All these observations add a few more dimensions to our experimental data and enhance our confidence in our results. Yes, we are expected to be objective. We have to know exactly what the accuracy, reproducibility, sensitivity, and specificity of an assay method are. We need instrumentation that converts the spots to peaks and numbers for numerical expression. But only those who work with QTLC know the degree of confidence obtained by the visual examination of the chromatogram, above and beyond the numerical criteria of reliability.

In a well-managed analytical laboratory, method selections are based on objective evaluation of needs, capabilities of methods, and available resources. QTLC is one of the indispensable techniques available to us to be used as an independent method or in combination with other methods complementing and enhancing one another.

Contributors

Daniel W. Armstrong Department of Chemistry and Biochemistry, Texas Tech University, Lubbock, Texas

F. Eisenbeiss Chemical Reagents Division, Research and Development Chromatography, E. Merck, Darmstadt, Federal Republic of Germany

H. E. Hauck Chemical Reagents Division, Research and Development Chromatography, E. Merck, Darmstadt, Federal Republic of Germany

Péter Horváth Applied Physical Chemistry Research Department, Chemical Works of Gedeon Richter, Ltd., Budapest, Hungary

Fred A. Huf Federal State Institute for Quality Control of Agricultural Products, Wageningen, The Netherlands

W. Jost Chemical Reagents Division, Research and Development Chromatography, E. Merck, Darmstadt, Federal Republic of Germany

Milica Katic Research Department, Lek Ljubljana, Ljubljana, Yugoslavia

K.-F. Krebs Chemical Reagents Division, Research and Development Chromatography, E. Merck, Darmstadt, Federal Republic of Germany

Franz Kreuzig Department of Bioanalytics, Research and Development, Biochemie GmbH, Kundl, Austria

Armin Medja Research Department, Lek Ljubljana, Ljubljana, Yugoslavia

Kumar D. Mukherjee Federal Center for Lipid Research, Münster, Federal Republic of Germany

Mirko Prosek Research Department, Lek Ljubljana, Ljubljana, Yugoslavia

Marko Pukl Department of Pharmacy, Edvard Kardelj University, Ljubljana, Yugoslavia

Gábor Szepesi Department of Quality Control and Quality Assurance, Chemical Works of Gedeon Richter, Ltd., Budapest, Hungary

John H. P. Tyman Department of Chemistry, Brunel, The University of West London, Uxbridge, Middlesex, England

Tom R. Watkins Department of Nutrition and Food Science, Hunter College, The City University of New York, New York, New York

Contents

Quantitative Thin-Layer Chromatography
and Its Industrial Applications

1

Prerequisites for Thin-Layer Chromatography in the Analytical Laboratory with Regard to Supplies and Materials

H. E. HAUCK, W. JOST, K.-F. KREBS, and F. EISENBEISS
Chemical Reagents Division, Research and Development Chromatography, E. Merck, Darmstadt, Federal Republic of Germany

I. INTRODUCTION

Thin-layer chromatography was accepted as a separation method in the analytical laboratory after suitable sorbents became available for self-preparation of thin-layer plates. This development took place at the beginning of the 1960s. Although even at this early stage a few manufacturers had standardized some parameters of the sorbents to a certain degree and the users were generally very skilled in the art of preparing thin-layer plates, the prepared plates could only be used for qualitative analysis. Because of the inhomogeneities in the packing structure, density, and surface structure of thin-layer plates coated by hand, quantitative analysis was not possible due to their inherently poor reproducibility and inadequate accuracy.

The desire to obtain high-quality thin-layer plates combined with improved reproducibility of the layer properties led to the development of machine-made precoated layers for thin-layer chromatography.

The invention of efficient instruments for sample application, development, and evaluation as one prerequisite for quantitative thin-layer chromatography (QTLC) led to an increase in the demands made on the quality of the precoated layers then available.

After these increased quality requirements had been met as a second prerequisite for QTLC, in particular by the development of precoated layers for high-performance thin-layer chromatography (HPTLC), thin-layer chromatography could be employed for the first time as a quantitative analytical method comparable with other such techniques (1-6).

II. DEFINITION OF SORBENT PARAMETERS WITH RESPECT TO QUANTITATIVE THIN-LAYER CHROMATOGRAPHY

A. Primary Parameters

The selectivity of a chromatographic system is determined chiefly by the chemical composition of the sorbent used, and particularly by its surface chemical properties. Its selectivity can be changed by varying the chemical structure of the entire skeleton of the sorbent and by specific chemical modification of its surface.

The chromatographic behavior of a sorbent is affected not only by these chemical parameters but also by physical factors. In the case of porous materials these factors are the specific surface area, the specific pore volume, the mean pore diameter, and the crystallographic structure.

1. Nonmodified Sorbents

Silica gels

Porous, surface-active silica gels are by far the most commonly used sorbents in quantitative thin-layer chromatography. These porous systems are therefore characterized by a comprehensive range of data.

The skeleton of silica gel sorbents is formed from SiO_4^{4-} tetrahedrons and displays an amorphous structure. The SiO_4^{4-} tetrahedrons at the surface of the sorbent are saturated by silanol groups (SiOH) or siloxane bridges (Si-O-Si).

In adsorption chromatography the surface silanol groups represent the active centers at which the interactions take place required for the retention of the compounds to be separated. These interactions are mainly hydrogen bonding, dipole-dipole interactions, and electrostatic interactions.

The density of silanol groups can be assumed to be constant with a value of approximately 8 $\mu mol/m^2$ for all types of silica gels used in QTLC (7). As the retention of sample substances in adsorption chromatography is determined largely by the number of accessible silanol groups, the strength of this retention is proportional to the specific surface area of the silica gel used.

The porous, surface-active silica gels employed by various manufacturers to prepare precoated layers suitable for QTLC have specific surface areas between 200 and 800 m^2/g. The main proportion of this area is formed by the surface within the pore system (internal surface area). The specific pore volumes of these silica gels lie within the range 0.5 to 2.0 ml/g.

In addition to the two structural parameters discussed so far—the specific surface area and the pore volume—the chromatographic result is also determined by the mean pore diameter. This is because the exchange processes between the mobile and stationary phases which lead to the separation of the sample substances take place at the surface or close to the surface of the stationary phase. The length of the pores (determined to some extent by the particle size) and particularly the pore diameter play a decisive role in determining the kinetics of these processes. The pore widths of the sorbents must be larger than the molecular dimensions of the largest substances to be separated in their solvated forms; only then can unimpeded access to the active centers of the internal surface be guaranteed.

The pore widths of the silica gels used for precoated layers are between 4 and 20 nm. The angstrom unit of length is often still used. In many cases these figures are employed in the designation of the type of sorbent.

The pore structure parameters mentioned above, including the specific surface area, are not independent of one another. The relationship between them can be described by a mathematical expression defined by Wheeler (8) which assumes that the pores are cylindrical in shape:

$$D = \frac{4 \times 10^4 V_p}{S_{BET}}$$

Table 1 Pore Structure Parameters for Silica Gels

Silica gel designation	Specific surface area (m^2/g)	Specific pore volume (m/g)	Calculated mean pore diameter (nm)
1	675	0.68	4
2	462	0.46	4
3	550	0.82	6
4	500	0.75	6
5	315	0.73	9
6	420	1.05	10

where D is the pore diameter (Å), V_p the specific pore volume (ml/g), and S_{BET} the specific surface area (m^2/g). The relationships between the pore structure parameters are illustrated in Table 1, which gives some examples of silica gels that are commonly used for QTLC.

The data compiled in Table 1 show that specification of a single parameter is not sufficient to characterize the pore structure; a satisfactory impression of the constitution of a silica gel pore system can be obtained only if all the parameters are taken into consideration.

In order to obtain reproducible retardation of the sample substances on precoated silica gel layers with a particular solvent, not only must the adsorbent used in adsorption chromatography be standardized, but also the state of activity of the silica gel must be controlled. Silica gels adsorb solvent molecules from the gas phase because of their surface activity. Some of the surface-active centers will be occupied by solvent molecules, especially if water is used; these sites will then no longer be available in their original form for interaction with the sample molecules to be separated. It is therefore advisable to determine the degree of activity of the silica gel (e.g., by adjusting it to a defined relative humidity) and thus influence the strength of retention of a given sample substance with a defined solvent.

This process is illustrated in Figure 1 by means of three chromatograms. They represent an adsorption chromatographic separa-

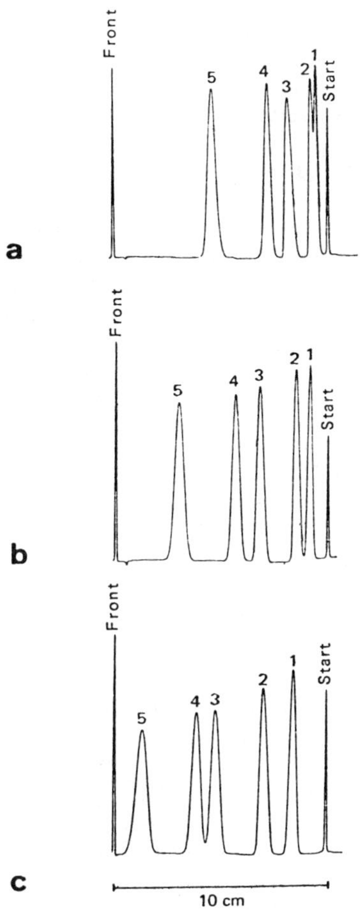

Figure 1 Effect of relative humidity on retention of lipophilic sample substances in adsorption chromatography. Plate: TLC silica gel 60 F254 precoated plate (E. Merck, Darmstadt, FRG). Substances: Ceres violet BRN (5), Ceres black G (4), Nitro fast blue 2B (3), Nitro fast violet FBL (2), Ceres red G (1). Eluent: toluene. Relative humidities (%): (a) 0, (b) 40, (c) 90. Type of chamber: Vario KS (Camag, Muttenz, Switzerland). Migration distance: 10 cm. Evaluation: in situ evaluation at 555 nm with a TLC/HPTLC scanner (Camag, Muttenz, Switzerland).

tion of five lipophilic dyes on a precoated silica gel 60 layer with toluene as the solvent at three different relative humidities (0, 40, and 90%) of the stationary phase. The defined relative humidity was adjusted via the gas phase by employing appropriate concentrations of sulfuric acid in a Vario KS chamber (9).

The strongest retentions can be observed for all the sample substances with 0% relative humidity and the weakest with 90% relative humidity. Furthermore, the separation of the strongly retarded substances improves as the relative humidity increases, but in the case of substances that are only slightly retarded the separation worsens as the silica gel is deactivated. This clearly shows that not only use of standardized sorbents but also control of the influence of the gas phase is important if reproducible results are to be obtained in QTLC.

Aluminum oxides

The aluminum oxides are the next most important surface-active sorbents after the silica gels. The aluminum oxides Al_2O_3 used for precoated layers in thin-layer chromatography are preferentially prepared from hydrargillite or γ-alumina. This starting product is thermally dehydrated and converted to γ- or χ-aluminum oxide or a mixture of the two which are suitable for chromatographic purposes. During dehydration the structural parameters which help to determine the chromatographic separation properties of these sorbents are fixed.

As is the case with the silica gels, the hydroxyl groups of the aluminum oxides that are located at the surface are primarily responsible for the retention interactions with sample molecules. Their surface density is approximately 13 $\mu mol/m^2$ (10). The specific surface areas of porous aluminum oxides employed as precoated layers are between 50 and 350 m^2/g; their specific pore volumes lie within the range 0.1 to 0.4 ml/g. The pore diameters derived from these structural parameters are 2 to 35 nm. The structural parameters for two commercially available aluminum oxide precoated layers are listed in Table 2.

Surface-active aluminum oxides, like silica gels, tend to adsorb water molecules from the surrounding atmosphere, leading to a reduction of their activity and their chromatographic retention capacity. Therefore it is advisable to standardize these surface activities to obtain reproducible chromatographic results. This can be done by using a procedure developed by Brockmann and Schodder (11), which allows classification of the aluminum oxide in question into five levels of activity, depending on its water content.

Table 2 Structural Parameters for Aluminum Oxide Precoated Layers

Al_2O_3 designation	Specific surface area (m^2/g)	Specific pore volume (ml/g)	Calculated mean pore diameter (nm)
1	200	0.3	6
2	70	0.2	11

Commercially available precoated aluminum oxide preparations usually have a pH value of about 9.0; however, precoated Al_2O_3 preparations are available that are adjusted to pH 7.0.

Kieselguhr and other similar compounds
Kieselguhr is a natural product of fossil origin that is formed from the skeletons of dead diatoms. These materials consist mainly of silicic acid and occur in deposits in various parts of the world. The chemical composition and physical properties of kieselguhrs vary as a result of their different biological origins and geological surroundings. Table 3 gives a summary of average values for the chemical composition of kieselguhrs that are used in the preparation of precoated layers for thin-layer chromatography.

Table 3 Chemical Composition of Kieselguhrs

Chemical component	%
SiO_2	90
Al_2O_3	2.5
Fe_2O_3	1.0
TiO_2	0.1
CaO	0.2
MgO	0.5
Na_2O, K_2O	1.0
H_2O, CO_2	4.7

(a)

Figure 2 Scanning electron micrographs of (a) kieselguhr and (b) silica 50,000. Magnification in both cases is ×3000.

Kieselguhrs must be carefully purified before being used to prepare precoated layers in order to remove the contaminants present as a result of their natural origin. These purification steps lead to changes in the original specific surface area and the pore structure. Purified kieselguhrs that can be employed for precoated plates have specific surface areas of about 1-4 m^2/g, specific pore volumes of about 1-3 ml/g, and mean pore widths of 1,000-10,000 nm. Purification does not, however, eliminate all the foreign atoms incorporated in the silicon dioxide skeleton. The remaining foreign atoms can affect the chromatographic behavior of the kieselguhr.

In addition to the naturally occurring kieselguhrs, a synthetically prepared inactive silicon dioxide (silica 50,000) with comparable chromatographic properties is used for precoated layers.

Figure 2 shows scanning electron micrographs (magnification ×3000) of kieselguhr (Figure 2a) and silica 50,000 (Figure 2b).

(b)

Figure 2 (Continued).

The original forms of the fossilized diatom skeletons can clearly be seen in the micrograph of the kieselguhr.

The highly pure silica 50,000 consists of 100% silicon dioxide and has a specific surface area of approximately 0.5 m^2/g, a specific pore volume of approximately 0.6 ml/g, and a mean pore width of 5000 nm. Both kieselguhr and silica 50,000 are mainly suitable for partition chromatography because of their low specific surface areas.

A special application of these inactive sorbents is their use in precoated layers with so-called concentrating zones (13). These precoated layers are composed of two different layer sections—the actual separating layer and the preceding concentrating zone. The separating layer consists of a surface-active adsorbent. The concentrating zone is composed of one of the two inactive sorbents and extends for 20 to 40 mm in the direction of chromatography, depending on the manufacturer.

When such precoated layers are used, the mixture of substances to be separated is applied to the concentrating zone. During development of this layer the substance spots are transported to the boundary between the concentrating zone and the separating layer by the ascending solvent and are concentrated here in a narrow band. The actual chromatographic separation process begins at this boundary. In this method chromatography and concentration are performed without interuption with the same solvent.

Precoated layers with concentrating zones offer a number of other advantages compared to layers without such zones:

Improvement of chromatographic separation efficiency, especially if large columes are applied.

Simple application technique; e.g., the starting spot does not have to be precisely positioned in the direction of chromatography.

Possible decomposition reactions of sensitive substances are avoided (contact of the substances with the surface-active adsorbent takes place only when they are dissolved in the solvent, analogous to wet application in liquid column chromatography).

Possibility of retardation of interfering accompanying substances (cleanup) in the concentrating zone.

Cellulose

The organic sorbent cellulose is employed quite often for precoated layers in thin-layer chromatography. Cellulose is a naturally occurring polysaccharide with the general formula $(C_6H_{10}O_5)_n$. Natural cellulose in the form of cellobiose units is purified prior to use in thin-layer chromatography and, if required, is subjected to crystallization.

The resulting native or microcrystalline celluloses for thin-layer chromatography differ with respect to the morphology of their particles. Native cellulose occurs in the form of long fibers; microcrystalline cellulose exists as rods.

In thin-layer chromatography both types of cellulose are used mainly to separate polar substances in partition chromatographic systems due to their relatively low surface areas (up to approximately 2 m^2/g).

Polyamides

Polyamides represent another type of organic sorbent, and they are also used for the preparation of precoated layers for thin-layer chromatography. There are two types: polyamide 6 and polyamide 11. Polyamide 6 is polycaprolactam and polyamide 11 is a polyundecanamide.

As polyamides have surface amide groups, they display a particularly high selectivity for substances that can form hydrogen bonds (e.g., phenols, carboxylic acids, amino acids).

2. *Surface-Modified Sorbents*

In addition to the nonmodified sorbents already discussed, silica gels and cellulose have been used as bases for preparing chemically modified phases in which the surface properties of the precoated layers for thin-layer chromatography are selectively changed.

Hydrophobically modified silica gels
In recent years the hydrophobic, so-called reversed phase (RP) precoated preparations in particular have become increasingly important (14-18). Silica gel with a mean pore width of preferably 6 to 10 nm is used as a base material for preparing hydrophobically modified precoated layers. Chemical modification is performed by reacting the accessible silanol groups with appropriate alkyl or aryl silanes. In this case silanol groups are eliminated and new siloxane groups are formed. The chain lengths of the alkyl silanes range from 1 to 18 carbon atoms.

The hydrophobic character of the modified silica gel increases with growing alkyl chain length and with the degree of modification. Consequently, hydrophobic materials of this type are no longer wetted by pure water or solvents that contain a high percentage of water, and it is no longer possible to develop chromatograms on these reversed phase precoated layers with these solvents. The reason for this behavior is that the capillary forces required for development are no longer large enough to overcome the hydrophobic repulsive forces.

Solvent systems with a high water content are often used in reversed phase chromatography because of the polarity of the substances to be separated. Reversed phase precoated layers with low defined degrees of modification but identical alkyl chain lengths have been developed so that reversed phase thin-layer chromatography is not limited from the outset (19, 20). These RP precoated layers display a hydrophobic but also a partially hydrophilic character as a result of the remaining accessible silanol groups; thus their use with solvents of different polarities is not restricted.

Hydrophilically modified silica gels
A hydrophilically modified precoated layer has also recently been used in thin-layer chromatography in addition to the nonpolar modified forms of silica gel (20-22). This material is an aminopropyl modification of silica gel. The selectivity of this functional

group is determined by its polar properties and its ability to act as a weakly basic ion exchanger.

Surface-modified cellulose
In the case of cellulose precoated layers it is also possible to realize surface modifications because of the presence of hydroxyl groups in this material. Acetylation of cellulose is one of the most important modifications in this respect. Precoated layers with different degrees of acetylation are available.

A second group of surface-modified cellulose precoated layers consists of weakly basic anion exchangers. The functional groups determining the selectivity are polyethylenimine (PEI), diethylaminoethyl (DEAE), and ECTEOLA (a reaction product of epichlorohydrin, triethanolamine, and alkali cellulose).

B. Secondary Parameters

The primary parameters discussed in the preceding sections—chemical composition, crystallographic structure, specific surface area, specific pore volume, and pore width—are not the only factors determining the chromatographic behavior of precoated layers. Attention should also be paid to secondary parameters when preparing precoated layers for QTLC, namely the mean particle size and the particle size distribution.

These two physical parameters are of decisive importance for the packing structure, surface homogeneity, migration behavior, and separation performance (but not for the selectivity) of precoated layers.

The sorbents used for precoated layers generally consist of irregular particles whose size cannot be described by a single specification of length. Different results are obtained, depending on the method employed to determine the particle size (e.g., sieving analysis, sedimentation, counting under the microscope) (23). Furthermore, the sorbents used in thin-layer chromatography display different distributions based on the size of the individual particles. Specification of a mean particle size is therefore not sufficient to characterize the dimensions of the sorbent; data on its particle size distribution are also required. Particle size distribution values (d_5, d_{10}, d_{50}, d_{90}, d_{95}) are usually determined by conventional particle analysis. The particle size distribution "d" values represent mass distributions for defined percentages.

The fine particles of the sorbent have a negative effect on the flow properties of precoated layers, and the large particles have an adverse effect on the spreading of the spot. Therefore a narrow particle size distribution is a decisive quality criterion.

In particular, the sorbents employed for high-performance thin-layer chromatography (HPTLC) satisfy these requirements to a high degree.

The mean particle size of the sorbent also has a decisive influence on the separation performance of precoated layers. Precoated layers with small mean particle diameters (HPTLC) display a higher separation efficiency than those with a larger mean particle diameter (TLC).

C. Demands on the Quality of Sorbents Used in the Preparation of Precoated Layers for QTLC

The primary and secondary parameters of the sorbents used for quantitative analysis in thin-layer chromatography must fulfill a series of quality criteria in order to guarantee accuracy and reproducibility. This applies especially to the quantitative in situ evaluation of thin-layer chromatograms. The most important demands made on the quality of a sorbent concern its chemical purity, the reproducibility and standardization of its pore structure, specific surface area, particle size distribution, and the reproducibility of the degree of modification in the case of chemically modified materials.

D. Demands on the Quality of Precoated Layers for QTLC

Precoated layers for QTLC are supplied on different supports, including glass plates, plastic foils, and aluminum foils.

One important criterion for glass plates used as supports for QTLC precoated plates is the accuracy of the measurements and angles of the specified format (e.g., 5 × 5, 5 × 10, 10 × 10, 10 × 20, 20 × 20 cm). The glass plates should also display no fluorescence; the factory fumes released during the production process should be removed as much as possible before the plate is coated.

Plastic foil supports for precoated layers should, above all, have sufficiently high stability and be resistant to the solvents employed in chromatography. As most plastic foils contain softeners and some low molecular weight polymers which can be dissolved by certain organic solvents, it is advisable to seal the foils with a suitable preliminary covering before they are actually coated.

If aluminum foils are used as supports for precoated layers, care must be taken that their surface is free from the grease employed for drawing during their manufacture; this grease would otherwise cause inhomogeneous coating.

A general quality criterion, which applies to all of the supports mentioned, is that the layer thickness should be as even and uniform as possible.

Precoated layers for QTLC must display sufficient mechanical stability to allow easy handling. This is achieved by incorporating a suitable binder or binding system in the chromatographic layer. The binding agent must satisfy the following requirements:

The binder should confer an adequate degree of solidity to the separating layer and allow it to adhere to the support.

The incorporated binder should not, if possible, affect the permeability of the separating layers.

The binder should not be attacked or eluted by the solvents commonly used in thin-layer chromatography.

The selectivity of the separating layer should be affected as little as possible by the binder.

The presence of the binder should not interfere with the detection of the sample substances in the separating layer. This applies both to the detection reagents used and to detection in the visible or UV range.

In order to obtain the best possible chromatographic separation performance in QTLC with precoated layers, suitable sorbents and above all an optimal packing structure are required. This means that the particles should be arranged as densely and as homogeneously as possible within the layer. An especially smooth, homogeneous layer surface can then be expected. Surfaces of this quality are absolutely necessary for precoated layers used in QTLC if low detection limits are to be obtained in the in situ evaluation of thin-layer chromatograms.

In addition to a homogeneous packing structure, a uniform layer thickness of the sorbent must be guaranteed if the requirements of QTLC are to be satisfied.

Fluorescence indicators have been incorporated into a series of precoated layers to facilitate the detection of certain sample substances. These indicators must be very homogeneously distributed within the layer, especially for evaluation of thin-layer chromatograms by the so-called fluorescence quenching method. The particle size distribution and the mean particle size of the indicator particles must be coordinated with the particle size of the sorbent used. Like the binders, the fluorescence indicators should not be eluted by the customary solvents.

Precoated layers for QTLC should be as pure as possible. This means that no visible contaminants should be present in the layer and that zones of impurities should not appear in the separation area or close to the solvent front after development.

III. SUMMARY

A number of manufacturers are endeavoring to meet the extensive demands that are made on sorbents and the precoated layers derived from them for QTLC. The use of thin-layer chromatography for quantitative analysis has been made possible by the development of such high-quality precoated layers.

REFERENCES

1. J. Ripphahn and H. Halpaap. J. Chromatogr. *112*, 81 (1975).
2. H. Halpaap and J. Ripphahn. Kontakte (Merck) *3/76*, 16 (1976).
3. U. Hezel, J. Blome, H. Halpaap, D. Jänchen, R. E. Kaiser, and J. Ripphahn. *Eine Einführung in die Hochleistungs-Dünnschichtchromatographie, HPDC*, Institute of Chromatography. Bad Dürkheim, Federal Republic of Germany, 1976.
4. A. Zlatkis and R. E. Kaiser. In: *HPTLC—High Performance Thin-Layer Chromatography*, Journal of Chromatoggraphy Library, vol. 9, Elsevier, Amsterdam, 1977.
5. H. Halpaap and J. Ripphahn. Chromatographia *10*, 613 (1977).
6. H. Halpaap and J. Ripphahn. Chromatographia *10*, 643 (1977).
7. H. P. Boehm. Angew. Chem. *78*, 617 (1966).
8. A. Wheeler. Catalysis, vol. 2, Reinhold, New York, 1955.
9. F. Geiss. *Die Parameter der Dünnschichtchromatographie*, Vieweg, Braunschweig, 1972.
10. C. Combellas and B. Drochon. Anal. Lett. *16* (A20), 1647 (1983).
11. H. Brockmann and H. Schodder. Berichte der deutschen chemischen gesellschaft, *74*, 73 (1941).
12. H. E. Hauck and H. Halpaap. Chromatographia *13*, 538 (1980).

13. H. Halpaap and K.-F. Krebs. J. Chromatogr. *142*, 823 (1977).

14. A. M. Siouffi and T. Wawrzynowics. J. Chromatogr. *186*, 563 (1979).

15. U. A. Th. Brinkmann and G. de Vries. J. High Resolut. Chromatogr. Chromatogr. Commun. *2*, 79 (1979).

16. H. Halpaap, K.-F. Krebs, and H. E. Hauck. J. High Resolut. Chromatogr. Chromatogr. Commun. *3*, 215 (1980).

17. U. A. Th. Brinkmann and G. de Vries. J. Chromatogr. *192*, 331 (1980).

18. U. A. Th. Brinkmann and G. de Vries. J. High Resolut. Chromatogr. Chromatogr. Commun. *5*, 476 (1982).

19. H. E. Hauck and W. Jost. Am. Lab. *15* (8), 72 (1983).

20. H. E. Hauck and W. Jost. GIT Fachz. Lab. Suppl. *3/83*, 3 (1983).

21. W. Jost and H. E. Hauck. J. Chromatogr. *261*, 235 (1983).

22. W. Jost and H. E. Hauck. Anal. Biochem. *135*, 120 (1983).

23. W. Reich. Kontakte (Merck) *3/77*, 26 (1977).

2

In Situ Evaluation of Thin-Layer Chromatograms

FRED A. HUF *Federal State Institute for Quality Control of Agricultural Products, Wageningen, The Netherlands*

I. INTRODUCTION

Quantitative analysis of mixture constituents often requires a preliminary separation. Thin-layer chromatography (TLC) has proved to be a relatively simple and inexpensive but also powerful method for the resolution of a great variety of mixtures (1-3). After separation, two different approaches for quantitation of the compounds are possible: extraction from the sorbent layer followed by some type of measurement (e.g., spectrophotometric and electrochemical) or measurements in situ.

Extraction of compounds from the sorbent layer has some drawbacks: it is time-consuming, relatively large amounts of substances are needed, and quantitative extractions are not always attainable. Direct evaluation of a thin-layer chromatogram is therefore attractive, and it can be considered a remarkable extension of TLC as an analytical method, as illustrated in other chapters of this book.

A simple but inaccurate way of direct evaluation is visual comparison of spot diameter and color intensity relative to reference spots. Such a procedure is subject to inaccuracy because human observations are imperfect. Accuracy is improved with instrumental methods. Several densitometers, instruments measuring light absorption and/or emission of the substances present in a chromatographic spot, are commercially available.

Since light absorption and emission are unambiguously related to the amount of substance present, densitometry offers a simple and rapid method for quantitation. The aim of this chapter is to consider theoretical aspects and to derive practical conclusions for successful application of the theoretical evaluations. It will be shown that commercially available densitometers are suitable instruments for obtaining reliable results under well-recognized circumstances. In the past disappointing results were obtained, and as a result densitometry was a rather unpopular technique. Poor analytical results were caused by irregularities of the sorbent layer thickness, which commercial densitometers cannot handle. Some manufacturers have now succeeded in fabricating TLC plates which are characterized by a constant layer thickness and homogeneous packing of the individual particles. As a consequence, the scattering power is constant over the plate area, yielding a better signal-to-noise ratio and an unmistakable relationship between light absorbance and substance concentration. The latter is strongly determined by the scattering power, as will be shown in this chapter. To a first approximation, densitometry with TLC plates of fluctuating layer thickness can be compared to classical spectrophotometry with cells of varying path length. Attention must be paid to this source of error when so-called home-made plates are measured. The errors can easily exceed 30% because such plates will usually show considerable, sometimes even visible, thickness variations.

Since the sorbent layer is light-scattering, the simple Lambert-Beer absorption law cannot be applied. This well-known combination of two laws, also referred to as the Bouguer-Beer law, states that when a monochromatic parallel light beam passes through a transparent medium, any layer of given thickness (dx) of the medium absorbs the same fraction of forward-traveling light incident on it. Absorption of radiant energy can be thought of as a result of collisions between photons and molecules, yielding (similar to a second-order reaction rate):

$$-\frac{di}{dx} = \varepsilon Ci \qquad [1]$$

where

ε = a proportionality factor called the molar absorption coefficient
C = the concentration of light-absorbing substance
i = the intensity of forward-traveling light, representing the number of photons passing layer dx per second

Solution of the differential equation [1] yields the well-known equation:

$$A = -\log T = \varepsilon CX \qquad [2]$$

where

A = absorbance
T = transmittance
X = layer thickness

The silica layer is not transparent but is a complex optical matrix. Incident radiation penetrates the layer and is partly reflected to the surface because of multiple scattering at the boundaries of the individual particles. The following notation will be used in this chapter (see Figure 1):

Is = radiant flux of incident light
Io = radiant flux transmitted by the blank layer
I = radiant flux transmitted by the TLC spot
Jo = radiant flux reflected by the blank layer
J = radiant flux reflected by the TLC spot

Different parameters, e.g., particle size, layer thickness, carrier of the sorbent, humidity, and acidity, may influence light absorption and light scattering. Several theories describing light absorption in scattering media have been developed (4). The theory usually applied to densitometry is that of Kubelka and Munk, which will be summarized in Section III. Before presenting this

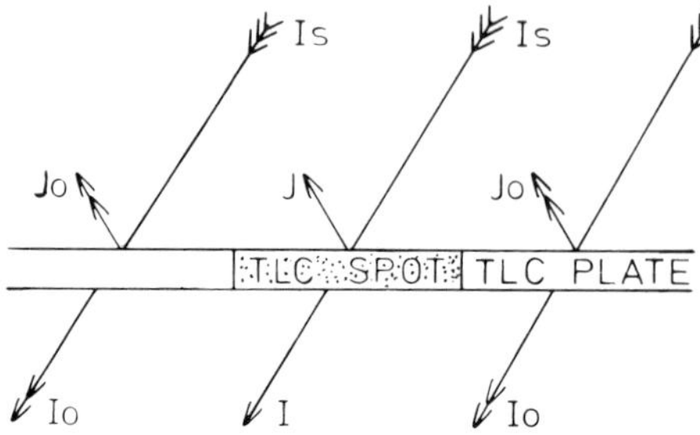

Figure 1 Light transport in a silica layer. Is, radiant flux of incident light; Io, radiant flux transmitted by the blank layer; I, radiant flux transmitted by the TLC spot; Jo, radiant flux reflected by the blank layer; J, radiant flux reflected by the TLC spot.

theory, it will be convenient to consider some basic concepts underlying the quantitation of thin-layer chromatograms, without regard to the assumed theory of light absorption.

II. PRINCIPLES OF IN SITU QUANTITATION OF CHROMATOGRAPHIC SPOTS

Generally, densitometric methods are based on nonlinear relationships between the substance concentration and the measured optical quantity q, where q is the measured value of transmittance and/or reflectance. Regardless of the kind of relationship used, the absorption coefficient K of the spot substance is correlated with any of the optical quantities by a function f:

$$K = f(q) \qquad [3]$$

The value of K, obtained after conversion of q according to Eq. [3], is related to the amount of substance in the TLC spot by a surface integral [5]. The substance mass dM in a surface element dy dz of the spot is given by (see Figure 2):

$$dM = cX\,dy\,dz \qquad [4]$$

where

c = substance concentration in the spot
X = layer thickness

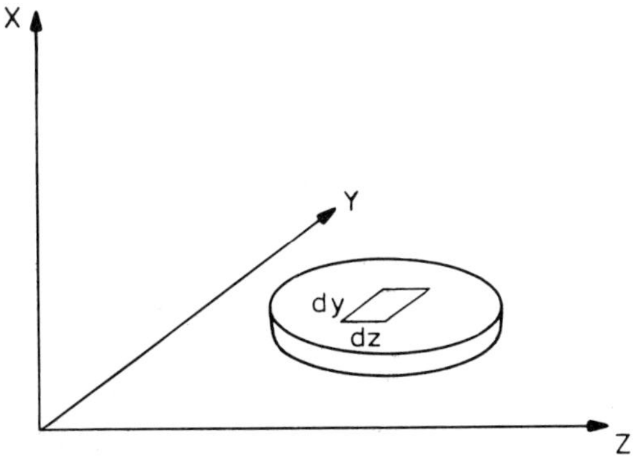

Figure 2 Integration of KX values over the spot area.

The absorption coefficient is proportional to the concentration:

$$K = \varepsilon c \qquad [5]$$

where

ε = molar absorption coefficient

Substituting Eq. [5] into Eq. [4] gives

$$dM = \frac{KX}{\varepsilon} dy\, dz$$

The total mass M is obtained after integration over the spot area:

$$M = \int_y \int_z \frac{KX}{\varepsilon} dy\, dz \qquad [6]$$

Obviously, a densitometer should scan the TLC spot with a light beam of small diameter (flying spot) and convert the value of q to KX, and during the scanning the KX value must be integrated continuously. After scanning, a densitometer response is obtained that is proportional to the amount of substance in the TLC spot, depending on the molar absorption coefficient and instrumental factors.

This procedure would be rather simple if the Lambert-Beer law was valid, e.g., after making the sorbent layer transparent by wetting. The desired KX value would then be found simply by

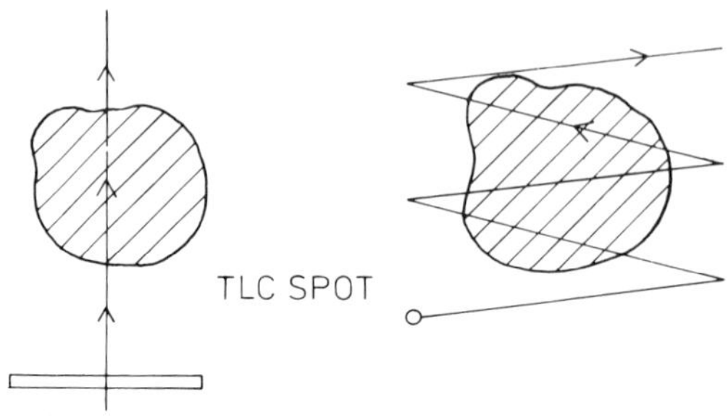

Figure 3 Scanning methods in commercial densitometers: (a) slit-scanning and (b) flying-spot scanning.

taking the negative logarithm of the transmittance during the scanning, according to Eq. [2]. A great deal of effort has been put in finding equations that are generally valid, as will be shown in Section V.

Most of the commercially available densitometers are provided with a so-called slit-scanning system instead of a flying-spot system (see Figure 3). A relatively large light beam is used to scan the TLC spot. The dimensions of the light beam exceed the diameter of the TLC spot. The advantages of this method are that apparatus construction is less complicated and optical noise (see Section VI) is averaged over the slit width. According to Eq. [6], the averaged KX value, $\overline{KX}$, is proportional to M, but $\overline{KX}$ is not simply correlated with the average q value. In general,

$$\overline{f(q)} \neq f(\overline{q}) \qquad [7]$$

because f is a nonlinear function of q, and at a given $\overline{KX}$ value the $\overline{q}$ value will therefore depend on the concentration distribution over the slit surface (6). The densitometer response is only comparable with chromatographic TLC spots when all the spots are of the same size and diameter and are positioned in a reproducible part of the scanning slit (7). This effect is probably caused by the angular distribution of the reflected diffuse light; in accordance with Lambert's cosine law, the detected fraction of the energy decreases to the outer part of the slit area and gives a decreasing contribution to the total light reflectance. Even when the spot is properly positioned, the densitometer response is still nonlinear with respect to M. As long as the curvature in the calibration curves is constant for all chromatograms, the nonlinearity is not an obstacle to quantification. However, this approach requires highly reproducible chromatographic procedures.

III. THEORY OF KUBELKA AND MUNK

The Kubelka-Munk (K-M) theory (8-10) is a relatively simple theory since all the matrix properties are represented by two parameters, K and S, the coefficients of absorption and scattering, respectively. The K-M theory describes light transmittance and reflectance as a function of both coefficients. The assumption that the absorption coefficient of the chromatographic substance is not influenced by the scattering properties of the silica layer, and vice versa, is of course a prerequisite for applying the K-M theory to densitometry. Furthermore, the medium is assumed to be a plane-parallel layer composed of light-absorbing and light-scattering particles whose dimensions are much smaller than the layer thickness X. To avoid edge effects, the layer is further assumed to have infinite lateral extension. The layer is illuminated with a source of monochromatic light.

The light is considered to travel in one direction, perpendicular to the plane of the layer. The radiant flux in the positive x direction is called i; the radiant flux in the negative direction is called j. The layer is then illuminated at the surface x = 0, and x = X is the unilluminated surface (see Figure 4).

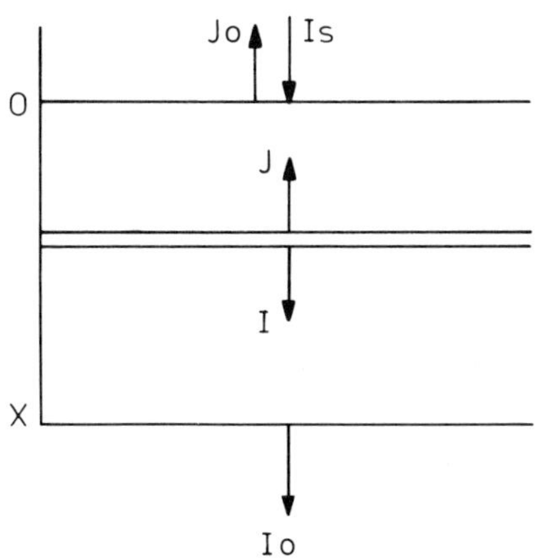

Figure 4 Light transport in a scattering medium according to the theory of Kubelka and Munk.

While passing a layer dx of the medium, the radiant flux will decrease because of absorption and scattering, and will increase because of scattering of the light traveling in the opposite direction.

$$-\frac{di}{dx} = (K + S)i - Sj \tag{8a}$$

$$\frac{dj}{dx} = (K + S)j - Si \tag{8b}$$

The fraction of light scattered in the opposite direction is assumed to be proportional to the radiant flux. The proportionality factor is called S, the scattering coefficient. The solution of the set of coupled differential equations [8], extensively discussed by Kortüm (11), yields:

$$T = \frac{I}{I_s} = \frac{b}{a\sinh(bSX) + b\cosh(bSX)} \tag{9}$$

$$R = \frac{J}{J_s} = \frac{1 - R_g[a - b\,\mathrm{cthg}(bSX)]}{a + b\,\mathrm{cthg}(bSX) - R_g} \tag{10}$$

where

$a = (S + K)/S$
$b = \sqrt{a^2 - 1}$
R_g = reflectance of the background (carrier) of the medium

The silica layer of TLC plates is generally supported by a glass plate or aluminum sheet. Assuming that $R_g = 0$ for glass, the reflectance of TLC plates on glass is, from Eq. [10],

$$R_{glass} = \frac{1}{a + b\,\mathrm{ctgh}(bSX)} \tag{11}$$

When aluminum is used as the support, it may be assumed that $R_g = 1$ as a rough approximation:

$$R_{A1} = \frac{1 - a + b\,\mathrm{ctgh}(bSX)}{a + b\,\mathrm{ctgh}(bSX) - 1} \tag{12}$$

The reflectance values for real supports are given at the end of Section VII. The values mentioned above will be used for the purpose of illustrating some aspects which are of practical importance. When light in the visible part of the spectrum is used, the silica layer will absorb hardly any radiant energy. Now $KX \cong 0$ and Eqs. [3] and [4] give the blank transmittance and blank reflectance. From Eq. [9]:

$$T = \frac{bSX/\sinh(bSX)}{aSX + bSX\,\mathrm{ctgh}(bSX)}$$

Series expansion of the hyperbolic forms gives:

$$\sinh bSX = bSX + \frac{(bSX)^3}{3} + \cdots$$

$$\mathrm{ctgh}\, bSX = \frac{1}{bSX} + \frac{bSX}{3} - \frac{(bSX)^3}{45} + \cdots$$

Substituting $KX \cong 0$ and neglecting second and higher powers of KX yields:

$$T = \frac{1}{asX + 1} \qquad [13]$$

Without absorption of radiant energy, the incident light is partly reflected because of internal scattering. The remaining part is transmitted (see Figure 5). When KX = 0, then a = 1, yielding:

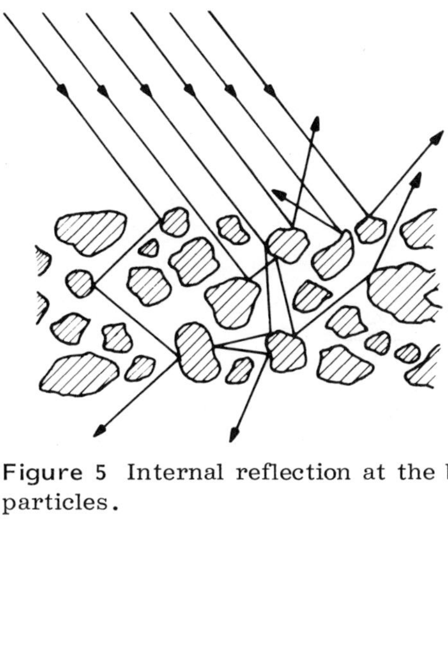

Figure 5 Internal reflection at the boundaries of the individual particles.

$$T_0 = \frac{1}{SX + 1} \tag{14}$$

Conversion of Eq. [11] in an analogous way yields:

$$R_{0,glass} = \frac{SX}{SX + 1} \tag{15}$$

The sum of transmittance and reflectance equals 1, in accordance with the fact that no radiant energy is absorbed.

The blank reflectance equals 1 for a TLC plate on a totally reflecting support. Because no absorption occurs, all the radiant energy is reflected:

$$R_{0,A1} = 1 \tag{16}$$

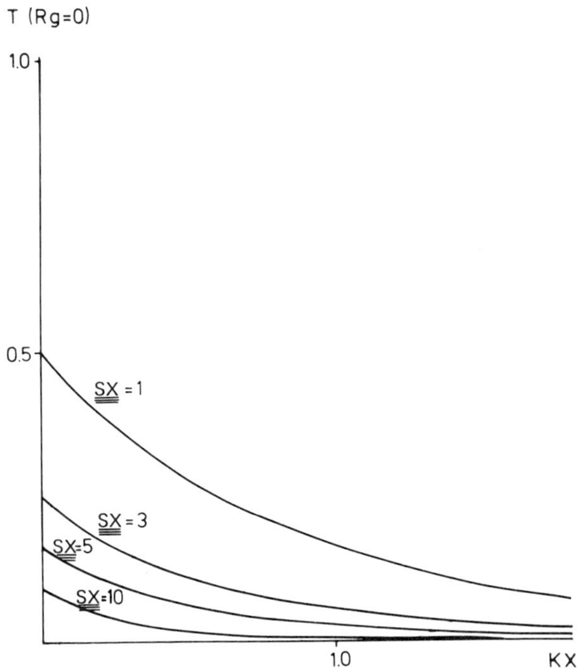

Figure 6 Transmittance values according to Eq. [9] at given SX values, assuming that the reflectance of the support is negligible: $R_g = 0$.

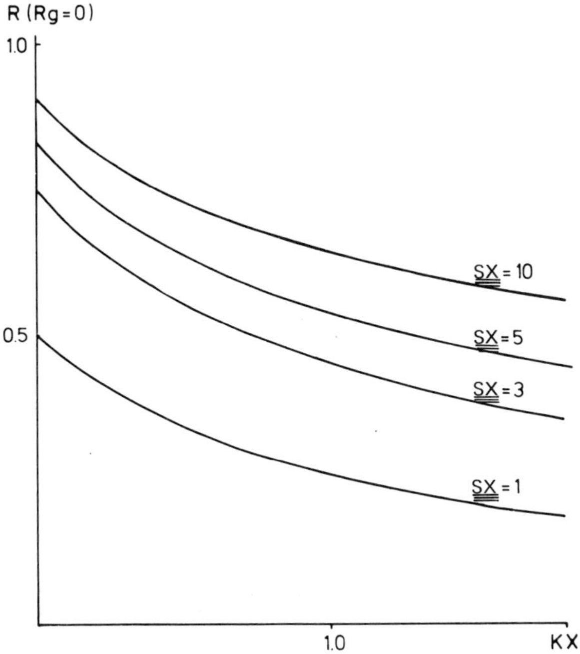

(a)

Figure 7 (a) Reflectance values according to Eq. [11] at given SX values, assuming that the reflectance of the support is negligible: $R_g = 0$. (b) Reflectance values according to Eq. [12] at given SX values, assuming a totally reflecting support: $R_g = 1$.

Equations [9], [11], and [12] are represented graphically in Figures 6 and 7 for some current SX values and the transmittance and reflectance values are also shown in Table 1.

Reflectance measurements are usually converted to KX with the well-known Kubelka-Munk remission function:

$$\frac{KX}{SX} = \frac{(1 - R_\infty)^2}{2R_\infty} \qquad [17]$$

This function is a particular solution of the set of differential equations when the layer thickness X is infinite. The layer thickness of a TLC plate is, of course, not infinite (the separation method is called *thin*-layer chromatography), but it may be

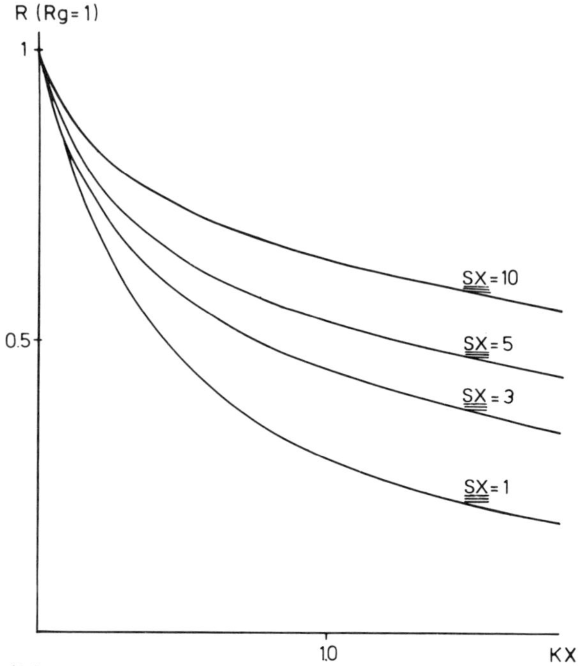

(b)

Figure 7 (Continued).

regarded as "apparently infinite," from an optical point of view, when bSX > 2 (12). (See also Section V.) Because:

$$\mathrm{ctgh}(bSX) = \frac{e^{bSX} + e^{-bSX}}{e^{bSX} - e^{-bSX}}$$

for bSX > 2 this yields:

$$\mathrm{ctgh}(bSX) = 1$$

and from Eq. [12]:

$$R = \frac{1 - R_g(a - b)}{a + b - R_g} \qquad [18]$$

Table 1 Transmittance and Reflectance Values Calculated According to the Theory of Kubelka and Munk ($1 \leqslant SX \leqslant 10$).[a]

SX = 1				SX = 2			
KX	T	RNUL	REEN	KX	T	RNUL	REEN
0	.50000	.50000	1.00000	0	.33333	.66667	1.00000
.05	.47176	.47986	.90774	.05	.31028	.64172	.91044
.10	.44528	.46103	.82892	.10	.28914	.61861	.83780
.15	.42044	.44341	.76101	.15	.26970	.59713	.77767
.20	.39712	.42689	.70206	.20	.25180	.57712	.72705
.25	.37520	.41138	.65055	.25	.23529	.55844	.68382
.30	.35460	.39681	.60528	.30	.22003	.54097	.64644
.35	.33523	.38311	.56528	.35	.20592	.52459	.61379
.40	.31699	.37020	.52975	.40	.19285	.50921	.58499
.45	.29982	.35803	.49806	.45	.18072	.49474	.55937
.50	.28365	.34655	.46967	.50	.16946	.48109	.53643
.55	.26840	.33569	.44414	.55	.15899	.46822	.51575
.60	.25403	.32543	.42109	.60	.14924	.45604	.49698
.65	.24048	.31571	.40022	.65	.14017	.44450	.47987

Table 1 (Continued).

SX = 1				SX = 2			
KX	T	RNUL	REEN	KX	T	RNUL	REEN
.70	.22769	.30650	.38125	.70	.13171	.43357	.46419
.75	.21562	.29776	.36397	.75	.12381	.42318	.44976
.80	.20422	.28947	.34816	.80	.11644	.41331	.43642
.85	.19345	.28159	.33368	.85	.10955	.40391	.42404
.90	.18328	.27409	.32036	.90	.10311	.39495	.41252
.95	.17367	.26695	.30809	.95	.09708	.38640	.40176
1.00	.16459	.26015	.29676	1.00	.09144	.37823	.39168
1.05	.15600	.25366	.28627	1.05	.08615	.37042	.38221
1.10	.14788	.24747	.27653	1.10	.08119	.36295	.37330
1.15	.14019	.24156	.26747	1.15	.07654	.35579	.36488
1.20	.13292	.23591	.25903	1.20	.07218	.34892	.35692
1.25	.12605	.23050	.25114	1.25	.06808	.34232	.34937
1.30	.11953	.22532	.24377	1.30	.06423	.33599	.34220
1.35	.11337	.22036	.23685	1.35	.06062	.32990	.33538
1.40	.10753	.21561	.23035	1.40	.05722	.32404	.32888

KX	T	RNUL	REEN	KX	T	RNUL	REEN
1.45	.10201	.21105	.22424	1.45	.05402	.31839	.32267
1.50	.09677	.20667	.21847	1.50	.05101	.31295	.31674
1.55	.09182	.20246	.21303	1.55	.04818	.30771	.31106
1.60	.08712	.19842	.20789	1.60	.04552	.30264	.30562
1.65	.08267	.19453	.20301	1.65	.04301	.29776	.30039
1.70	.07845	.19079	.19839	1.70	.04065	.29303	.29537
1.75	.07445	.18718	.19400	1.75	.03842	.28846	.29054
1.80	.07066	.18371	.18982	1.80	.03632	.28404	.28588
1.85	.06707	.18036	.18584	1.85	.03434	.27976	.28140
1.90	.06366	.17712	.18205	1.90	.03248	.27562	.27707
1.95	.06043	.17400	.17843	1.95	.03072	.27160	.27290
2.00	.05737	.17999	.17496	2.00	.02906	.26771	.26886

SX = 3				SX = 4			
KX	T	RNUL	REEN	KX	T	RNUL	REEN
0	.25000	.75000	1.00000	0	.20000	.80000	1.00000
.05	.22929	.72309	.91296	.05	.18066	.77210	.91530
.10	.21072	.69480	.84562	.10	.16368	.74676	.85256
.15	.19399	.67566	.79168	.15	.14870	.72362	.80363

Table 1 (Continued).

SX = 3				SX = 4			
KX	T	RNUL	REEN	KX	T	RNUL	REEN
.20	.17888	.65463	.74728	.20	.13542	.70238	.76400
.25	.16520	.63511	.70990	.25	.12360	.68278	.73095
.30	.15277	.61695	.67788	.30	.11304	.66463	.70273
.35	.14146	.59999	.65001	.35	.10358	.64774	.67820
.40	.13114	.58411	.62547	.40	.09507	.63199	.65654
.45	.12170	.56921	.60360	.45	.08739	.61723	.63718
.50	.11306	.55520	.58394	.50	.08045	.60337	.61969
.55	.10514	.54198	.56612	.55	.07416	.59032	.60374
.60	.09785	.52950	.54985	.60	.06844	.57800	.58910
.65	.09115	.51768	.53490	.65	.06324	.56634	.57556
.70	.08497	.50647	.52110	.70	.05850	.55528	.56298
.75	.07927	.49582	.50829	.75	.05417	.54478	.55123
.80	.07400	.48569	.49634	.80	.05021	.53478	.54020
.85	.06913	.47604	.48516	.85	.04658	.52524	.52981
.90	.06462	.46683	.47466	.90	.04324	.51613	.52000

.95	.06043	.45803	.46476	.95	.04018	.50742	.51069
1.00	.05655	.44960	.45542	1.00	.03736	.49907	.50186
1.05	.05295	.44154	.44656	1.05	.03477	.49106	.49344
1.10	.04960	.43380	.43815	1.10	.03237	.48337	.48540
1.15	.04648	.42637	.43014	1.15	.03016	.47598	.47772
1.20	.04358	.41924	.42251	1.20	.02812	.46886	.47035
1.25	.04088	.41237	.41521	1.25	.02623	.46200	.46328
1.30	.03836	.40576	.40823	1.30	.02449	.45539	.45649
1.35	.03601	.39938	.40154	1.35	.02287	.44900	.44995
1.40	.03381	.39324	.39512	1.40	.02136	.44283	.44365
1.45	.03176	.38730	.38895	1.45	.01997	.43686	.43757
1.50	.02985	.38157	.38301	1.50	.01867	.43109	.43170
1.55	.02806	.37602	.37729	1.55	.01747	.42549	.42602
1.60	.02638	.37066	.37177	1.60	.01635	.42007	.42053
1.65	.02482	.36547	.36644	1.65	.01531	.41431	.41321
1.70	.02335	.36044	.36129	1.70	.01434	.40970	.41005
1.75	.02197	.35556	.35631	1.75	.01344	.40474	.40505
1.80	.02068	.35083	.35149	1.80	.01260	.39992	.40019
1.85	.01948	.34624	.34682	1.85	.01181	.39524	.39547

Table 1 (Continued).

SX = 3				SX = 4			
KX	T	RNUL	REEN	KX	T	RNUL	REEN
1.90	.01834	.34178	.34229	1.90	.01108	.39068	.39088
1.95	.01728	.33745	.33790	1.95	.01040	.38624	.38642
2.00	.01628	.33323	.33363	2.00	.00976	.38192	.38208

SX = 5				SX = 6			
KX	T	RNUL	REEN	KX	T	RNUL	REEN
0	.16667	.83333	1.00000	0	.14286	.85714	1.00000
.05	.14824	.80488	.91750	.05	.12511	.82836	.91956
.10	.13242	.77930	.85875	.10	.11019	.80277	.86433
.15	.11872	.75615	.81395	.15	.09752	.77977	.82296
.20	.10680	.73503	.77808	.20	.08668	.75894	.79011
.25	.09637	.71565	.74831	.25	.07734	.73992	.76292
.30	.08719	.69778	.72293	.30	.06924	.72244	.73971
.35	.07907	.68122	.70083	.35	.06218	.70629	.71945
.40	.07187	.66579	.68125	.40	.05599	.69128	.70144

45.	.06546	.65138	.66368	.45	.05055	.67728	.68520
.50	.05974	.63787	.64772	.50	.04574	.66416	.67039
.55	.05460	.62515	.63310	.55	.04147	.65183	.65677
.60	.04999	.51315	.61961	.60	.03768	.64020	.64414
.65	.04584	.60179	.60707	.65	.03430	.62919	.63236
.70	.04209	.59103	.59536	.70	.03127	.61875	.62131
.75	.03869	.58079	.58436	.75	.02855	.60882	.61091
.80	.03562	.57104	.57400	.80	.02610	.59936	.60106
.85	.03282	.56174	.56420	.85	.02390	.59033	.59172
.90	.03027	.55285	.55490	.90	.02191	.58269	.58283
.95	.02795	.54434	.54605	.95	.02011	.57341	.57435
1.00	.02584	.53617	.53761	1.00	.01848	.56546	.56624
1.05	.02390	.52833	.52954	1.05	.01700	.55782	.55847
1.10	.02213	.52079	.52182	1.10	.01565	.55046	.55101
1.15	.02050	.52354	.51440	1.15	.01442	.54338	.54383
1.20	.01901	.50654	.50727	1.20	.01331	.53654	.53692
1.25	.01764	.49979	.50041	1.25	.01228	.52994	.53026
1.30	.01638	.49327	.49380	1.30	.01135	.52355	.52382
1.35	.01522	.48697	.48742	1.35	.01049	.51737	.51760

Table 1 (Continued).

SX = 5				SX = 6			
KX	T	RNUL	REEN	KX	T	RNUL	REEN
1.40	.01415	.48088	.48126	1.40	.00971	.51139	.51158
1.45	.01316	.47497	.47530	1.45	.00899	.50558	.50575
1.50	.01225	.46925	.46953	1.50	.0083	.49995	.50009
1.55	.01141	.46370	.46394	1.55	.00773	.49449	.49460
1.60	.01063	.45831	.45852	1.60	.00717	.48917	.48927
1.65	.00991	.45307	.45325	1.65	.00665	.48401	.48409
1.70	.00924	.44799	.44814	1.70	.00618	.47898	.47905
1.75	.00862	.44304	.44317	1.75	.00574	.47409	.47415
1.80	.00805	.43823	.43834	1.80	.00534	.46932	.46937
1.85	.00752	.43354	.43364	1.85	.00497	.46467	.46472
1.90	.00702	.42897	.42906	1.90	.00462	.46014	.46018
1.95	.00657	.42454	.42460	1.95	.00430	.45572	.45576
2.00	.00614	.42018	.42025	2.00	.00401	.45141	.45144

	SX = 7			SX = 8		
KX	T	RNUL	REEN	T	RNUL	REEN
0	.12500	.87500	1.00000	.11111	.88889	1.00000
.05	.10779	.84604	.92150	.09434	.85983	.92332
.10	.09362	.82053	.86937	.08081	.83449	.87395
.15	.08180	.79780	.83090	.06974	.81207	.83795
.20	.07186	.77733	.80052	.06057	.79199	.80963
.25	.06342	.75872	.77539	.05289	.77381	.78618
.30	.05621	.74168	.75391	.04642	.75721	.76609
.35	.05000	.72596	.73509	.04091	.74194	.74843
.40	.04462	.71139	.71829	.03620	.72780	.73262
.45	.03994	.69782	.70310	.03214	.71464	.71826
.50	.03585	.68511	.68919	.02862	.70233	.70508
.55	.03226	.67316	.67635	.02556	.69076	.69287
.60	.02909	.66190	.66440	.02289	.67984	.68148
.65	.02629	.65124	.65322	.02055	.66952	.67079
.70	.02380	.64112	.64270	.01848	.65972	.66072
.75	.02159	.63150	.63277	.01666	.65039	.65119

Table 1 (Continued).

SX = 7				SX = 8			
KX	T	RNUL	REEN	KX	T	RNUL	REEN
.80	.01962	.62233	.62335	.80	.01505	.64150	.64213
.85	.01785	.61357	.61439	.85	.01361	.63299	.63350
.90	.01627	.60518	.60585	.90	.01234	.62484	.62525
.95	.01485	.59714	.59768	.95	.01120	.61703	.61735
1.00	.01357	.58941	.58986	1.00	.01018	.60951	.60978
1.05	.01241	.58187	.58234	1.05	.00926	.60228	.60249
1.10	.01137	.57481	.57512	1.10	.00844	.59530	.59548
1.15	.01043	.56791	.56816	1.15	.00770	.58857	.58872
1.20	.00957	.56123	.56144	1.20	.00703	.58207	.58218
1.25	.00870	.55479	.55490	1.25	.00643	.57577	.57587
1.30	.00808	.54855	.54869	1.30	.00589	.56968	.56976
1.35	.00744	.54250	.54262	1.35	.00540	.56377	.56383
1.40	.00685	.53664	.53674	1.40	.00495	.55803	.55809
1.45	.00632	.53095	.53104	1.45	.00454	.55246	.55251
1.50	.00583	.52543	.52550	1.50	.00417	.54705	.54709

1.55	.00538	.52006	.52012	1.55	.00384	.54179	.54182
1.60	.00497	.51484	.51489	1.60	.00353	.53667	.53669
1.65	.00460	.50976	.50980	1.65	.00325	.53168	.53170
1.70	.00425	.50481	.50485	1.70	.00300	.52681	.52683
1.75	.00394	.49999	.50002	1.75	.00276	.52207	.52209
1.80	.00365	.49529	.49532	1.80	.00255	.51745	.51746
1.85	.00338	.49070	.49073	1.85	.00235	.51293	.51294
1.90	.00313	.48623	.48625	1.90	.00217	.50852	.50853
1.95	.00291	.48186	.48187	1.95	.00201	.50421	.50422
2.00	.00270	.47759	.47760	2.00	.00186	.50000	.50000

SX = 9				SX = 10			
KX	T	RNUL	REEN	KX	T	RNUL	REEN
0	.10000	.90000	1.00000	0	.09091	.90909	1.00000
.05	.08361	.87090	.92505	.05	.07485	.87999	.92667
.10	.07065	.84577	.87814	.10	.06240	.85510	.88197
.15	.06022	.82369	.84427	.15	.05256	.8338	.84995
.20	.05172	.80402	.81767	.20	.04465	.81410	.82483
.25	.04471	.78627	.79562	.25	.03822	.79678	.80396

Table 1 (Continued).

SX = 9				SX = 10			
KX	T	RNUL	REEN	KX	T	RNUL	REEN
.30	.03887	.77011	.77668	.30	.03291	.78101	.78598
.35	.03395	.75526	.75997	.35	.02851	.76660	.77008
.40	.02979	.74154	.74497	.40	.02481	.75327	.75576
.45	.02624	.72877	.73131	.45	.02169	.74087	.74268
.50	.02320	.71683	.71873	.50	.01904	.72928	.73062
.55	.02057	.70561	.70705	.55	.01677	.71839	.71939
.60	.01830	.69503	.69613	.60	.01481	.70812	.70887
.65	.01632	.68502	.68586	.65	.01313	.69840	.69897
.70	.01459	.67551	.67617	.70	.01167	.68916	.68960
.75	.01307	.66646	.66697	.75	.01039	.68037	.68071
.80	.01173	.65782	.65823	.80	.00928	.67197	.67224
.85	.01056	.64956	.64988	.85	.00830	.66394	.66414
.90	.00951	.64164	.64190	.90	.00744	.65623	.65639
.95	.00859	.63404	.63424	.95	.00669	.64883	.64896
1.00	.00777	.62673	.62689	1.00	.00602	.64170	.64180

1.05	.00703	.61968	.61981	1.05	.00542	.63484	.63492
1.10	.00638	.61289	.61299	1.10	.00489	.62821	.62827
1.15	.00579	.60632	.60641	1.15	.00442	.62180	.62185
1.20	.00527	.59997	.60004	1.20	.00401	.61650	.61565
1.25	.00479	.59383	.59389	1.25	.00363	.60960	.60963
1.30	.00437	.58787	.58792	1.30	.00329	.60378	.60380
1.35	.00399	.58210	.58213	1.35	.00299	.59813	.59815
1.40	.00364	.57649	.57652	1.40	.00272	.59264	.59265
1.45	.00333	.57103	.57106	1.45	.00248	.58730	.58731
1.50	.00305	.56573	.56576	1.50	.00226	.58210	.58212
1.55	.00279	.56057	.56059	1.55	.00279	.56057	.56059
1.60	.00256	.55555	.55557	1.60	.00256	.55555	.55557
1.65	.00235	.55065	.55067	1.65	.00235	.55065	.55067
1.70	.00215	.54588	.54589	1.70	.00215	.54588	.54589
1.75	.00198	.54122	.54123	1.75	.00198	.54122	.54123
1.80	.00182	.53667	.53668	1.80	.00182	.53557	.53668
1.85	.00167	.53223	.53224	1.85	.00167	.53223	.53224
1.90	.00154	.52789	.52789	1.90	.00154	.52789	.52789

Table 1 (Continued).

SX = 9				SX = 10			
KX	T	RNUL	REEN	KX	T	RNUL	REEN
1.95	.00142	.52364	.52365	1.95	.00142	.52364	.52365
2.00	.00131	.51949	.51950	2.00	.00131	.51949	.51950

[a]REEN is the reflectance with $R_g = 1$ and RNUL with $R_g = 0$.

From Eq. [17]:

$$R_\infty = \left(1 + \frac{KX}{SX}\right) - \sqrt{\left(\frac{KX}{SX}\right)^2 + 2\,\frac{KX}{SX}} \qquad [19]$$

$$= a - b \qquad [20]$$

$$= \frac{1}{a + b} \qquad [21]$$

Substitution of Eqs. [20] and [21] into Eq. [18] yields:

$$R = \frac{1 - R_g R_\infty}{1/R_\infty - R_g}$$

and

$$R = R_\infty$$

Obviously, the background of the medium is of no importance any longer. Only a negligible part of the reflected light penetrates the layer as far as the background, and the layer thickness can be regarded as infinite.

Transmittance and/or reflectance have some distinct advantages for the quantitative evaluation of thin-layer chromatograms. Reflectance is, of course often the only possible measuring mode when radiant energy is poorly transmitted (UV part of the spectrum) or not transmitted at all (aluminum support), but it may be a less sensitive method (13). Table 1 shows that transmittance gives a better response then reflectance for SX values around 3. Equations [14] and [15] are very suitable for determining the SX value of a TLC plate, and this value is found for several commercially available plates. The reflectance response depends on R_g and increases with increasing value. As a consequence, carriers with a strongly reflecting surface are preferable in reflectance densitometry. A comparison of reflectance and transmittance densitometry is given by Frei (14).

The signal-to-noise ratio will often be the most important parameter for comparing analytical methods (detection limit!), and will be treated in Section VI.

IV. ANGULAR DISTRIBUTION OF THE TRAVELING LIGHT

The assumption that the light travels only in a direction perpendicular to the plane of the layer is not realistic. Incident light is scattered because of internal reflections in the sorbent layer. Generally, Lambert's cosine law is assumed to describe the angular distribution of diffuse scattered light. This is a phenomenologically formulated law, based on the fact that a mat surface appears uniformly lucid at all angles of observation (see Figure 8).

When θ is the angle of observation with respect to the surface normal, then the radiant energy detected must be constant for all values of θ. The observed area of the mat surface increases with $\cos\theta$. In order to detect a constant level of light energy, the radiant flux per unit of solid angle, emitted by an observed surface element (dO), must decrease with $\cos\theta$ (15). The indicatrix of the radiant flux is then represented by a sphere (see Figure 9).

It was recognized by Kubelka (9) that different components of diffuse light passing a layer dx of the medium are not absorbed in an equivalent way. At an angle θ, the path length is not dx, but $dx/\cos\theta$ (see Figure 10). The path length increases at increasing θ values, the light absorption is correspondingly increased. The effect of the angular light distribution on absorption of the traveling flux is found by integrating the individual absorbance of every component over all values of θ. This integration is simplified by imagining a surface element dO' on a sphere around dO. From Lambert's cosine law (see Figure 11),

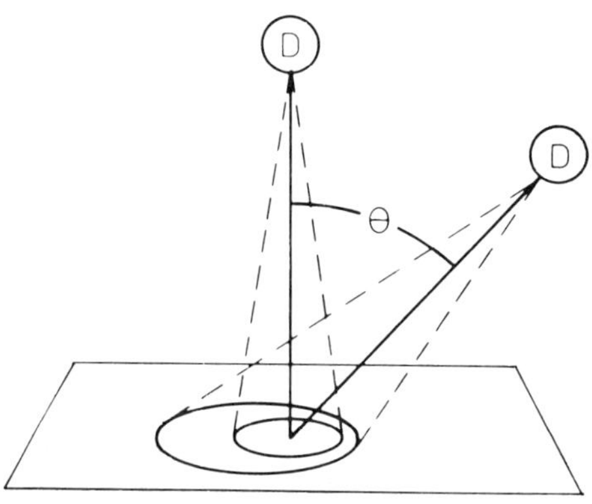

Figure 8 Observation of a mat surface at varying angle with respect to the surface normal (D, detector).

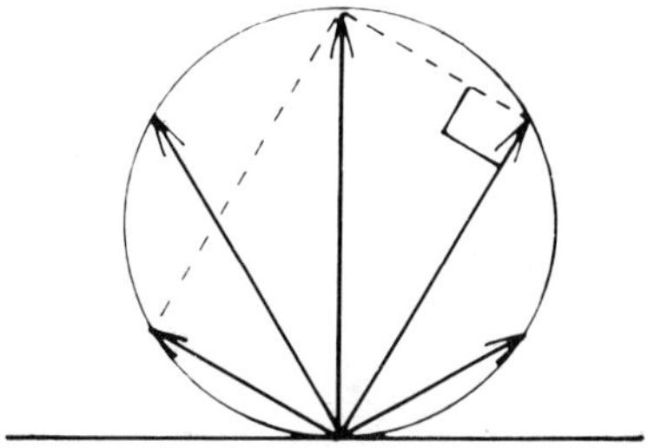

Figure 9 Indicatrix of diffuse scattered light.

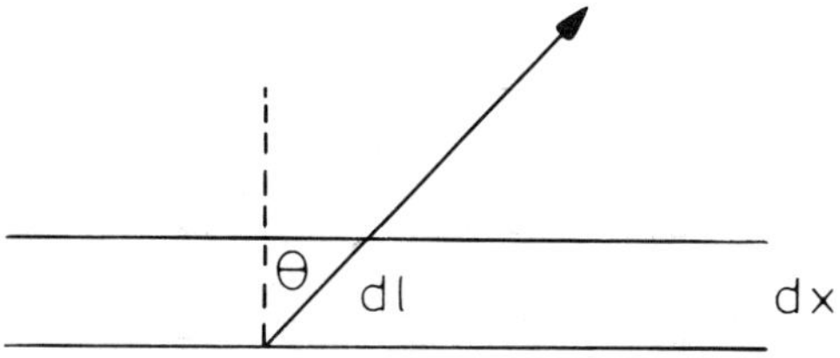

Figure 10 Path length of traveling light when passing a layer dx at an angle θ with respect to the normal.

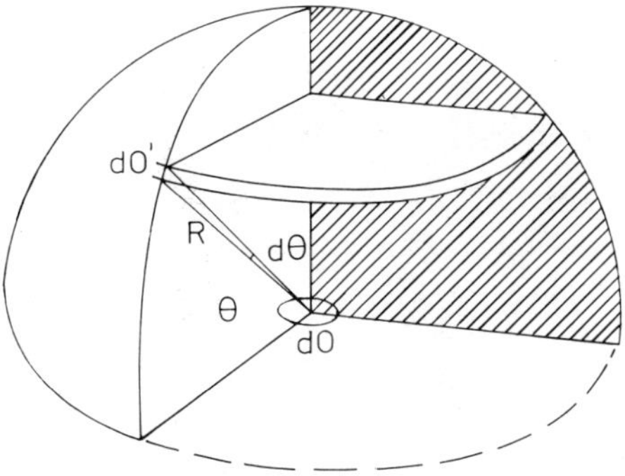

Figure 11 Light absorbance as a function of the angular light distribution.

$$dF = B \cos dO\, dO'/R^2 \qquad [22]$$

where:

B = proportionality factor
dF = fraction of radiant energy emitted by dO and passing dO'
dO' = surface element between angles θ and $(\theta + d\theta)$ on a sphere with radius R around dO
$dO' = 2\pi R^2 \sin\theta\, d\theta$

This yields, with Eq. [22],

$$dF = \pi B\, dO \sin 2\theta\, d\theta \qquad [23]$$

All energy F emitted by dO is found after integration of dF over all relevant values of θ:

$$F = \int_0^{\pi/2} B\, dO \sin 2\theta\, d\theta$$

$$F = \pi B\, dO$$

This yields, with Eq. [23],

$$dF = F \sin 2\theta\, d\theta \qquad [24]$$

When passing a layer dx, dF will diminish due to absorption:

$$-d(dF) = K\, dF \frac{\sin 2\theta}{\cos\theta} d\theta \qquad [25]$$

The total decrease in the radiant energy is found after integration over all values of θ:

$$-\frac{dF}{dx} = KF \int_0^{\pi/2} \frac{\sin 2\theta}{\cos\theta} d\theta$$

$$-\frac{dF}{dx} = 2KF$$

Apparently the decrease of radiant flux for diffuse scattered light, when passing through a layer dx, is twice the decrease of radiant flux for a parallel light beam of equal energy. Light absorption depends on the angular distribution of the traveling light and may be better described by a so-called effective absorption coefficient, the value of which is 2K for diffuse scattered light, where K is the well-known absorption coefficient in "classical" spectrophotometry.

Illumination of a TLC plate with a source of collimated light may, therefore, introduce errors in quantitative evaluation of the chromatogram. The incident light is scattered more and becomes more diffuse depending on the penetration depth. When the light is traveling through the medium, the effective absorption coefficient will vary from K to 2K. After it leaves the medium, the effective coefficient will have some average value, between K and 2K, depending on the layer thickness. When the layer thickness of the TLC plate fluctuates, absorption measurements are not directly comparable for different chromatographic spots. This type of error is eliminated, of course, by using plates of constant layer thickness or by using a source of diffuse scattered light in the densitometer. In this case, the effective coefficient will be constant for all measurements. To a first approximation a parallel light beam may be used as an alternative light source for illumination, when the light source is mounted in such a way that the incident light strikes the TLC plate at an angle of 60° with the surface normal of the plate. The effective absorption coefficient then equals 2K, just as for diffuse light, and this value is maintained during scattering. This method of illumination seems to be a simple solution, especially for reflectance measurements. For convenience, it is assumed in this chapter that diffuse light is used.

V. APPLICATION OF THE KUBELKA-MUNK THEORY TO DENSITOMETRY

Transmittance and reflectance are presented in Eq. [9] and [10] as a function of KX, SX, and R_g. Surprisingly, the effect of R_g on transmittance is not described in Eq. [9], and it will be dealt with at the end of this section. Since KX equals the mass of substance per unit of surface, the surface integral of KX equals the mass of substance in this chromatographic spot, as shown from Eq. [6]. The equations derived in the K-M theory must, therefore, be inverted when applied to densitometry, to obtain KX as a function of the measured transmittance and/or reflectance. The inverse functions will be nonlinear, as may be concluded from Figures 6 and 7, and will depend on SX, since the KX value at a particular reflectance or transmittance is strongly determined by the SX value of the medium under consideration, as also illustrated in these figures. Such an approach was chosen by Goldman and Goodall (13) for transmittance measurements, and they succeeded in deriving an inverse function of Eq. [9]. In spite of its relatively simple form, inversion of this equation is not (yet) performed in a mathematically analytical way. For that reason, Goldman and Goodall chose a graphical approximation. It is assumed that KX can be well approximated by two independent functions of the blank and spot absorbance, A_0 and A, respectively:

$$0.434KX = f(A_0)g(A) \tag{27}$$

where:

$$f(A_0) = \left(\frac{\delta KX}{\delta A}\right)_{A=0}$$

and

$$g(a) = A + \alpha A^2$$

The function $f(A_0)$ is determined by the scattering properties of the sorbent layer as follows from Eq. [14], $A_0 = \log(SX + 1)$, and is used as a correction for the hyperchromic effect caused by scattering. Because of internal reflections, the path length of the transmitted light increases, resulting in increased absorption. The function g(A) is used to linearize the calibration curves representing absorbance versus KX.

Differentiating Eq. [9] yields:

$$f(A_0) = \frac{3}{1 + 2\cosh(2.3A_0)} \tag{28}$$

Plotting $\log f(A_0)$ versus A_0 shows that $f(A_0)$ approximates $2\exp(-2A_0)$ very well in the range $0.7 < A_0 < 1.3$.

A numerical comparison of g(A) with the transmittance values from the Kubelka-Munk theory yields $\alpha = 0.4$ in the range $0 < A < 1$. For this value of α, Eq. [27] deviates only up to 3% from Eq. [9].

Substitution of $f(A_0)$ and α in Eq. [27] yields:

$$0.434KX = 2e^{-2A_0}(A + 0.4A^2) \tag{29}$$

Table 2 Values of the Coefficients in Eq. [30] for Varying SX and $0.45 < A_0 < 0.7$

SX	A_0	α	β	γ
2-5	0.45-0.7	$0.35A_0$-0.06	1.55	1.67

Goldman and Goodall used an impractical A_0 range in the derivation of Eq. [29]. Most prefabricated TLC plates appear to have A_0 values in the range $0.45 < A_0 < 0.7$, and Eq. [27] can be evaluated for general applicability to densitometry (16):

$$0.434KX = \beta - \exp(-\gamma A_0)(A + \alpha A^2) \qquad [30]$$

which is valid for $A < 1$. The coefficients in the range $0.45 < A_0 < 0.7$ are presented in Table 2.

It seems obvious to apply the same procedure for reflectance measurements, since up to the present no inverse function of Eq. [10] has been described. By analogy to Eq. [27], the following is obtained:

$$0.434KX = f(A_0)g(A) \qquad [31]$$

with

$$A_0 = -\log R_0$$

and

$$A = -\log R$$

$f(A_0)$ and $g(A)$ are assumed to be independent functions:

$$F(A_0) = \left(\frac{\delta KX}{\delta A}\right)_{A=0}$$

Then:

$$\frac{1}{f(A_0)} = \left(\frac{\delta A}{\delta KX}\right)_{KX=0}$$

From the definition of A,

$$\frac{1}{f(A_0)} = \frac{R}{2.3} \left| \frac{\delta(1/R)}{\delta KX} \right|_{KX=0}$$

and from Eq. [11]:

$$\frac{1}{f(A_0)} = \left| \frac{R}{2.3} \left\{ \frac{\delta a}{\delta KX} + \frac{\delta b}{\delta KX} \text{ ctgh}(bSX) + b \frac{\delta[\text{ctgh}(bSX)]}{\delta KX} \right\} \right|_{KX=0} \qquad [32]$$

Series expansion of the hyprebolic forms and neglecting the second and higher powers of KX yield:

$$\text{ctgh}\, bSX = \left(\frac{1}{bSX} + \frac{bSX}{3} - \frac{(bSX)^3}{4S} + \cdots \right)$$

$$= \frac{1}{bSX} - \frac{bSX}{3} \qquad [33a]$$

$$\text{sinh}\, bSX = \left(bSX + \frac{(bSX)^3}{3!} + \frac{(bSX)^5}{5!} + \cdots \right) = bSX \qquad [33b]$$

Substitution of Eqs. [33a] and [33b] into Eq. [32] yields:

$$\frac{1}{f(A_0)} = \left[\frac{R}{2.3} \left| \frac{1}{SX} + \frac{a}{bSX} \left(\frac{1}{bSX} + \frac{bSX}{3} \right) - \frac{a}{(bSX)^2} \right| \right]_{KX=0}$$

When KX → 0, then a → 1, b → 0, and R → SX/(SX + 1):

$$\frac{1}{f(A_0)} = \left\{ \frac{SX}{2.3(SX+1)} \left| \frac{1}{SX} + \frac{a}{3} + a \left[\frac{1}{(bSX)^2} - \frac{1}{(bSX)^2} \right] \right| \right\}_{KX=0}$$

$$\frac{1}{f(A_0)} = \frac{3 + SX}{2.3(3SX + 3)}$$

and

$$f(A_0) = \frac{2.3}{1 - 2/3\, R_0} \qquad [34]$$

Substituting Eq. [34] into Eq. [31] and equating Eqs. [31] and [11] yield the α term. For varying KX and SX values, α is graphically presented in Figure 12.

Figure 12 shows that α cannot be substituted by a single value in Eq. (30). Further evaluation of g(A) is now needed, including third and higher powers of A:

$$g(A) = A + \alpha A^2 + \beta A^3 + \cdots$$

The solution of the inversion has then lost its simplicity and is not applied to reflectance measurements.

Reflectance measurements are often converted to KX with the Kubelka-Munk remission function presented in Eq. [17] (1, 17-20). The remission function is a very simple relationship that can be applied directly to densitometry, provided the sorbent layer behaves as an infinite layer. Generally, regression lines will be obtained with a negative intercept, as can be shown by substitution of R from Eq. [12] into Eq. [17] for matched KX values, because

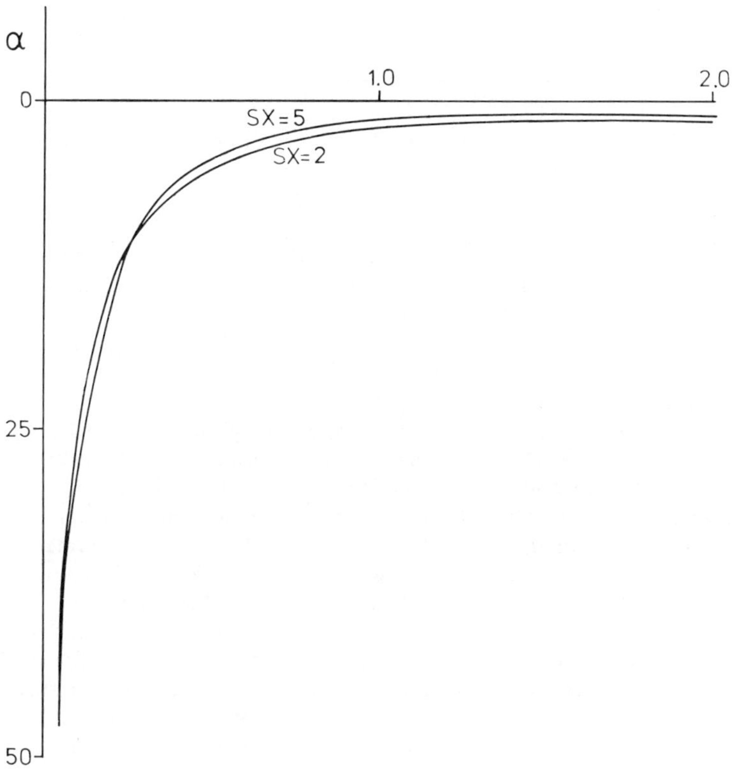

Figure 12 Dependence of α on KX at given SX values.

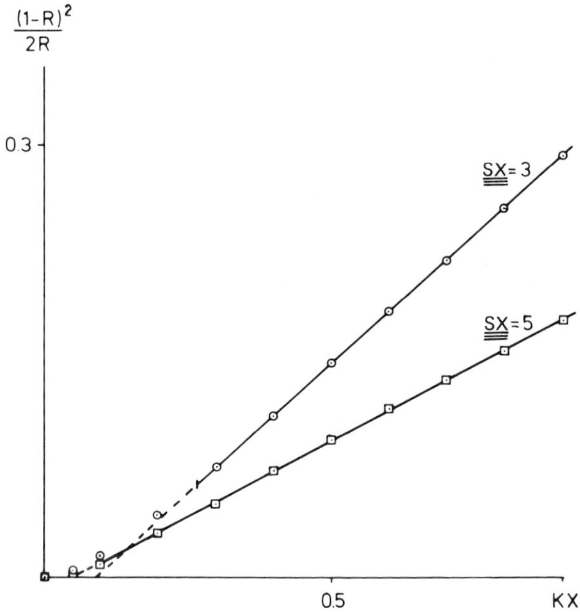

Figure 13 Regression lines obtained by applying the remission functions to reflectance measurements.

the remission function is not valid at low KX values. Calibration curves for some SX values are given in Figure 13. At higher substance concentrations and/or measurements with UV radiation, the medium can actually be considered infinite. Table 1 shows the range of KX in which the effect of R_g has disappeared, and this offers a simple but adequate criterion.

The effects of R_g will disappear in UV absorbance measurements since the blank layer absorbs radiant energy, as illustrated in the photoacoustic spectrum of silica gel (see Fig. 14).

Measurements with UV light are often performed at 254 nm, an intensive Hg emission line. At this wavelength the reflectance of a 1-cm silica layer is 0.64, as compared to a 1-cm MgO layer, yielding a ratio K/S = 0.1 (21). Since MgO slightly absorbs radiation, it is not totally reflecting, and the ratio will exceed 0.1. Current commercial plates (SX = 3) will therefore yield quantitative results with an error less than about 10% when the remission function is applied to UV measurements. The reflectance must, however, be corrected for layer thickness fluctuations, as illustrated by the following examples.

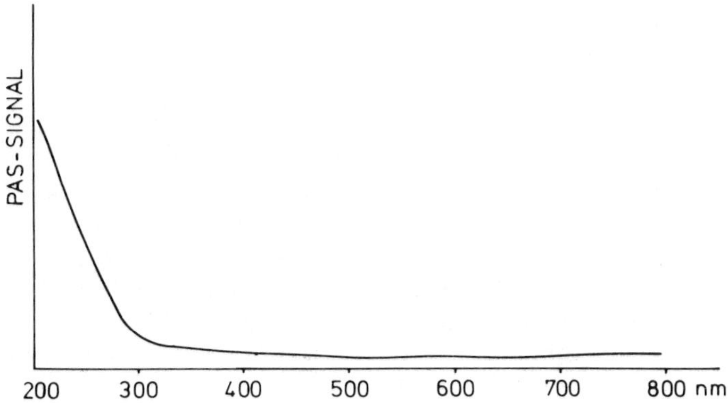

Figure 14 Photoacoustic spectrum of silica gel E. Merck H 60.

Suppose that two spots of equal concentration and diameter are measured (Figure 15). The apparatus response is equal for both spots, whereas the substance mass is proportional to the layer thickness. For that reason, the densitometer response needs continuous correction of the reflectance with X, e.g., via SX. If such correction is not possible, this error source can, of course, be excluded by the use of TLC plates with constant layer thickness.

Pollak and Boulton (22) derived an inverse function of the K-M equation by means of a transmittance line model. Although this approach has several interesting aspects, the derived function is, in the end, identical to the remission function, as can be proved by substitution of the hyperbolic term in the K-M equation by exponential terms:

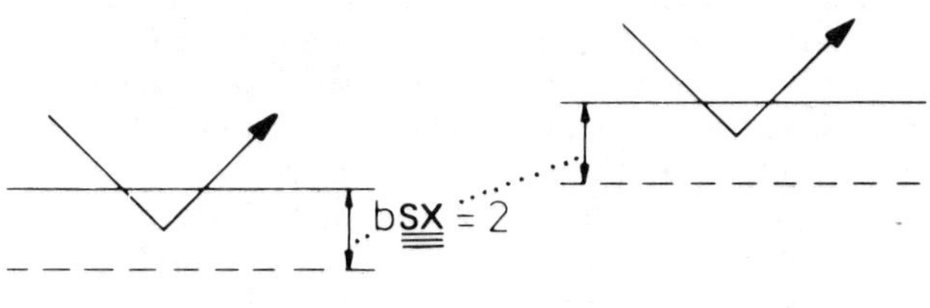

Figure 15 Effect of layer thickness variations when reflectance densitometry is performed on TLC plates with apparently infinite layer thickness.

$$\text{cotgh} = \frac{e^{bSX} + e^{-bSX}}{e^{bSX} - e^{bSX}}$$

To our knowledge, no mathematical inversion that is generally valid has (yet) been realized. An electronic inversion is, therefore, obvious and is introduced by Shimadzu in the CS densitometers. The photomultiplier signal is fed into a "linearizer," and amplifier with a gain characteristic that compensates the curvature of the absorption curves in the reflectance mode. The gain characteristic is adjusted by a set of potentiometers for several SX values. The theoretical background (23) is not carefully evaluated in these instruments, but good analytical results are obtained with an empirical adjustment of linearizers. A numerical inversion with a digital dedicated system is described by Huf (24). The values in Table 2 are stored in an EPROM. After they are digitized, the reflectance values are addressed to one of the eight pages of the EPROM (eight different pages are available for different SX values), and the corresponding KX data are fed into a digital-to-analog converter. This type of signal conversion is, of course, very flexible and can also be used for any measuring mode with complex mathematical relationships.

Transmittance and reflectance can also be measured simultaneously during the scanning of the chromatographic spot. Simultaneous measurement as a densitometric method is described by Treiber and co-workers (20, 25, 26) and Jork (19). Transmittance values are converted according to Lambert's absorption law, which is valid at relatively low substance concentrations (the term involving the second power of A in Eq. [30] is negligible compared to A). This procedure, proposed by Treiber, improves the performance of the remission function by combining it with the Lambert-Beer law for transmittance. The linearity of the densitometer response in Treiber's communicaiton is satisfying, but the calibration curves do not always pass through the origin. This deviation is probably imputable to insufficient correction of the remission function at low substance concentrations. Additional critical notes about the procedure followed are given by Treiber (26). In the procedure proposed by Treiber (25) and Jork (19), reflectance and transmittance are added. Baseline fluctuations and optical noise in the densitogram are eliminated when measurements are performed with light in the visible part of the spectrum, from Eqs. [14] and [15] it is shown that $T_0 + R_0 = 1$, regardless of layer thickness fluctuations. Little attention is paid to the linearity of the densitometer response. It can be shown that the densitometer response is linear over a limited range of low concentrations (12). This procedure is therefore very suitable at low concentrations with slit-scanning instruments.

The K-M equations can be combined and inverted in a mathematical way for simultaneous measurement. The following result is obtained (12). From Eq. [9]:

$$T^2 = \frac{b^2}{\sinh^2 bSX\ (a + b\,\mathrm{ctgh}\, bSX)}$$

and with:

$$\mathrm{ctgh}\, bSX = \frac{(1 + \sinh^2 bSX)}{\sinh^2 bSX}$$

this yields after substitution of Eq. [11]:

$$T^2 = R^2 - 2aR + 1$$

and

$$a = \frac{R^2 + 1 - T^2}{2R}$$

or

$$KX = SX\, \frac{(1 - R + T)(1 - R - T)}{2R} \qquad [35]$$

Equation [35] offers a simple relationship for densitometric application, because no hyperbolic functions are present and optical noise and baseline fluctuations are eliminated. When measurements are performed with visible light, $T_0 + R_0 = 1$ and the apparatus response equals zero. A constant baseline is obtained when the silica layer shows considerable absorption of radiant energy, assuming a constant layer thickness. The ratio KX/SX can be regarded as a material constant, the value of which causes an offset of the baseline. In case of a fluctuating layer thickness, the instrument response is simply corrected by multiplicaiton by SX, the value of which is obtained, e.g., from the ratio

$$SX = R_0/T_0$$

For an (apparently) infinite layer thickness, the transmitted radiant energy is negligible compared to the reflected radiant energy. Then $T << R$, and Eq. [35] reduces to the previously mentioned remission function.

Finally, it should be noted that absolute values of transmittance and/or reflectance are necessary for application of the K-M equations, as can be illustrated by some straightforward calculations with Eq. [17]. The absorbance mode is much more convenient in this respect since the ratio of (relative) light intensities of the blank layer and the chromatographic spot is measured. Absolute measurements are subject to practical hindrances, particularly for reflectance, since it is difficult to find a good relaible reference for absorption measurements in scattering media (11). It has been shown that the remission function is suitable for obtaining the absolute reflectance of silica sorbent layers (24). The experimental work is, however, time-consuming and small measuring errors can cause considerable aberrations in the reflectance values obtained. The assumptions that $R_g = 0$ for a glass plate and $R_{Al} = 1$ for an aluminum sheet are convenient in manipulating Eq. [10] but are not correct: for the glass and the aluminum sheets of E. Merck H 60 plates, $R_g = 0.14$ and $R_{Al} = 0.69$, respectively. The effect of R_g on both transmittance and reflectance will therefore be considered (24). In the absence of a carrier, the reflectance R_f is represented by Eq. [11]. Substituion of R_f in Eq. [10] yields

$$R = \frac{R_f - R_g(2aR_f - 1)}{1 - R_f R_g}$$

With Eq. [35] this yields

$$R = \frac{R_f - R_g R_f^2 + R_g T_f^2}{1 - R_f R_g}$$

or

$$R = R_f + \frac{R_g T_f^2}{1 - R_f R_g} \qquad [36]$$

Obviously, the reflectance of the free layer is increased by a fraction ΔR when a carrier is added:

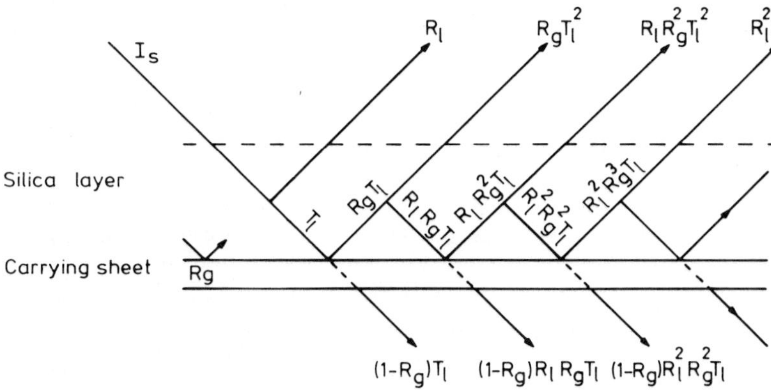

Figure 16 Effect of carrier on light transmittance and reflectance.

$$\Delta R = \frac{R_g T_f^2}{1 - R_f R_g}$$

The fraction ΔR is the result of partial reflection of the transmitted radiant energy when it reaches the boundary of the medium at the side of the carrier, as illustrated in Figure 16. The reflected part is again involved in the scattering process. Figure 16 also shows that the transmitted light that actually passes the carrying sheet, and will be measured, decreases by reflection at the boundary:

$$T = T_f(1 - R_g) + (1 - R_g)T_f[1 + R_f R_g + (R_f R_g)^2 + \cdots]$$

$$T = T_f \left(\frac{1 - R_g}{1 - R_f R_g} \right) \qquad [37]$$

As mentioned before, the effect of R_g on transmittance is not considered in the K-M equation. Probably, Kubelka and Munk's main interest did not include radiant transmittance, since they were involved with "Farb-anstriche."

Transmittance will decrease by about 6% relatively (SX = 3) and reflectance will increase by about 1% relatively when a glass plate is used as a carrier, and the equations derived above will hold to a good approximation. The effect of aluminum sheets is restricted to reflectance only. Equation [12] proves to be an unacceptably rough approximation, and for this type of TLC plates, Eq. [10] must be rewritten as follows:

$$R_{Al} = \frac{1 - 0.69(a - b\,ctgh\,bSX)}{a + b\,ctgh\,bSX - 0.69} \qquad [38]$$

The curvature of the calibration lines is therefore different from those obtained with glass plates (see also Figure 6), introducing a source of error in the straightforward application of the Shimadzu densitometers. The R_g values mentioned were obtained for E. Merck H 60 plates. It is obvious that before spectrophotometric evaluation, the R_g and SX values of the TLC plates developed must be carefully determined.

VI. SIGNAL-TO-NOISE RATIO

High-power light sources and "high-performance" electronic equipment make electrical noise almost arbitrarily small. Baseline noise in the densitogram is, therefore, restricted to optical noise, which is mainly caused by thickness fluctuations of the sorbent layer. Further contributions to optical noise may be caused by nonuniform particle size, local changes in composition, etc. Inhomogeneous distribution of the fluorescence indicator introduces considerable extra nosie. The best signal-to-noise (S/N) ratio is generally obtained for fluorodensitometry, because the blank layer does not show interfering fluorescence. A good (S/N) ratio is also obtained for simultaneous T, R densitometry or measurement at two or more wavelengths, because optical noise can then be effectively eliminated. Comparison of transmittance and reflectance in this respect, for convenience restricted to the absorption of visible light, requires some evaluation of Eqs. [13], [14], and [15].

Optical noise N introduced by variations or irregularities of the blank sorbent layer may be defined as:

$$N_T = \left(\frac{\delta T}{\delta X}\right)\Delta X = \left(\frac{\delta T}{\delta SX}\right)\Delta SX \qquad N_R = \left(\frac{\delta R}{\delta X}\right)\Delta X = \left(\frac{\delta R}{\delta SX}\right)\Delta SX$$

for transmittance and reflectance, respectively (27). This yields in Eqs. [14] and [15]:

$$N_T = -\frac{\Delta SX}{(SX + 1)^2} \qquad N_R = \frac{\Delta SX}{(SX + 1)^2}$$

Optical noise is equal in magnitude but opposite in sign for the two methods (for this reason optical noise in simultaneous reflectance and transmittance densitometry is eliminated). The signal-to-noise ratio q can be compared with Eq. [13], which is valid for low substance concentrations. The S/N ratio is of practical importance at low substance concentrations only (detection limit) and for low KX values:

$$q_T = \frac{(SX + 1)^2}{(SX + KX + 1)} \frac{\Delta KX}{\Delta SX} \qquad [39]$$

and

$$q_R = \frac{SX(SX + 1)^2}{KX + SX + 1} \frac{\Delta KX}{\Delta SX} \qquad [40]$$

Comparison of Eqs. [39] and [40] shows that reflectance measurements have a signal-to-noise ratio that is a factor SX better than that of transmittance measurements.

The better signal-to-nosie ratio in reflectance can be further improved by an increased R_g value (24). Expanding the hyperbolic form in Eq. [10] in a series and neglecting second and higher powers yield:

$$R = \frac{SX(1 - R_g) + R_g - R_g KX}{SX(1 - R_g) + 1 + KX}$$

so that

$$q_R = \frac{SX(1 - R_g)(1 + R_g) + 2R_g}{(1 - R_g)(1 + KX - R_g + R_g KX)}$$

which can be rewritten as

$$q_R = \frac{SX[(1 + R_g/(1 - R_g)] + 2R_g/(1 - R_g)}{1 - R_g(1 - KX) + KX}$$

The S/N ratio obviously increases with increasing R_g value, yielding an interesting limiting condition: for KX = 0 and $R_g \to 1$, $S/N \to \infty$ and apparently the optical noise is completely eliminated. Both conditions are not met in practice, but it can be concluded that, in general, the best signal-to-noise ratio is obtained in the reflectance mode for a sorbent layer on a support with a high R_g value. For practical purposes, an aluminum sheet is to be preferred over a glass plate as a support for this reason, and also because of its better sensitivity, as mentioned before. Of course, the aluminum sheet must be absolutely flat during measurements in order to exclude variations in the detected fraction of the diffuse scattered light (e.g., see Figure 9). The equations derived for the S/N ratio are valid only for radiation in the visible part of the spectrum. As the wavelength decreases the sorbent layer will show increasing absorbance, yielding an increasing bSX value.

The effect of R_g decreases with increasing bSX and disappears for bSX > 2, as follows from Eq. [17]. Reflectance of the blank layer is then dependent only on the ratio K/S, which can be regarded to a first approximation as a constant for a particular silica gel layer and a particular wavelength. Layer thickness variation will not cause optical noise, but it must be absent in order to avoid erroneous measurements, as illustrated in Figure 15.

VII. FLUORODENSITOMETRY

Although densitometry refers to methods dealing with the measurement of optical density, hence light absorbance, it is usual to include fluorometric methods as well for the evaluation of thin-layer chromatograms. This measuring mode is usually referred to as fluorodensitometry, and it is an attractive method for quantitation, since silica gel does not show fluorescence in the part of the spectrum currently applied. In order to collect all the fluorescence radiation, a feflecting support, e.g., an aluminum sheet, is preferable. By far the most frequently applied excitation wavelengths are the Hg emission lines at 254 and 366 nm. A densitometric relationship is easily derived from the discussion in the previous sections. The silica layer shows considerable absorbance at these wavelengths, as illustrated in Figure 14, and for that reason the remission function can be applied to a good approximation.

To another good approximation, the intensity F of the emitted radiation is directly proportional to the fraction of radiant energy absorbed by the analyte:

$$F = \frac{K}{K_S + K}\,\theta(1 - R)I_S \qquad [42]$$

where

θ = quantum efficiency
K_S = absorption coefficient of silica gel

From the remission function

$$1 - R = \frac{KX + K_S X}{SX}\;\frac{2R}{1 - R}$$

This yields with Eq. [42]:

$$KX = SX\,\frac{F}{\theta I_S}\left(\frac{1 - R}{2R}\right) \qquad [43]$$

For low KX values, Eq. [43] can be rewritten as follows:

$$\Delta KX = SX\,\frac{F}{\theta I_S}\;\frac{1}{\sqrt{[1 + 2SX/(K_S X + \Delta KX)]} - 1} \qquad [44]$$

This equation shows a fluorescence intensity that is directly proportional to the concentration for the range in which ΔKX is small compared to $K_S X$, that is, at low concentration levels. This is hardly surprising, since a similar situation is known from fluorometry of solutions. The range in which a (less sensitive) linear response is obtained increases with increasing absorption coefficient of the sorbent layer. Eventually, a correction for nonlinearity at higher concentration levels can be realized, in accordance with Eq. [43], via simultaneous measurement of the reflectance at the excitation wavelength, provided the change in reflectance is measured with sufficient accuracy.

Fluorescence quenching is a technique that is widely used in qualitative analysis of UV-absorbing substances. A nonmigrating substance, called a fluorescence indicator, e.g., zinc silicate with an excitation wavelength at 254 nm, is thoroughly mixed with the sorbent layer. The fluorescence intensity diminishes when a substance present absorbs incident light, since less radiant energy is available for the indicator. Fluorescence diminishing, rather than quenching, would, therefore, be a better description for this type of measurement. Only a minor fraction of the incident energy is absorbed by the indicator and, analogously to Eq. [42],

$$F = \frac{K_i}{K_i + K_S + K} \theta_i (1 - R) I_S \qquad [45]$$

where

K_i = absorption coefficient of the indicator
θ_i = quantum efficiency of the indicator

Since the quenching Q is the relative decrease in blank fluorescence F_0, from Eq. [45]:

$$Q = \frac{F_0 - F}{F_0} = \frac{\sqrt{1 + 2SX/(K_S X + K_i X)} - \sqrt{1 + 2SX/(K_S X + K_i X + KX)}}{\sqrt{1 + 2SX/(K_S + K_i)} - 1} \qquad [46]$$

The denominator of the right-hand part of Eq. [5] is a matrix constant, and from the numberator it is obvious that the response of Q to variation of KX is rather poor, making fluorescence quenching the least sensitive densitometric method. Furthermore, the indicator proves to be irregularly distributed over the sorbent layer, causing considerable optical noise. As a result, this measuring mode is certainly the least attractive one, but it can be improved by covering the TLC plate with a transparent homogeneous sheet in which an indicator is present. Sheets with different indicators would make quenching more versatile. Optical noise is also reduced when the indicator is supplied after development of the thin-layer plates, e.g., by dipping.

VIII. QUANTITATION IN HIGH-PERFORMANCE TLC

HPTLC has proved to be a rapid and powerful separation method, the success of which is due in large part to the manufacturer who succeeded in producing TLC plates with silica gel in a well-defined small particle range and with a constant layer thickness (3).

Because of the relatively small particle diameter, HPTLC plates show an SX value around 7 when measured with visible radiation. Because of the increased scattering power, the resolution of the scanning system in the densitometer cannot be below 0.2 mm.

In HPTLC, paltes are developed over a short distance (several centimeters) and separation power is fully utilized only when the chromatographic spots are very small. As a result, quantitative evaluation in accordance with Eq. [6] is not possible, since this equation requires a resolution that is much smaller than the diameter of the chromatographic spot. Even the smallest possible beam diameter of incident light in a flying-spot densitometer will, because of scattering, yield results that are similar to those obtained with slit-scanning instruments. A slit-scanning device is preferable under these circumstances because the eventual optical noise is averaged over the slit area and the instrument can be less expensive because its construction is much simpler. As mentioned above, very good reproducibility in the chromatographic procedure is a first requirement when slit-scanners are used, and peripheral equipment for this purpose is commercially available (28). With the exception of fluorodensitometry at low concentration levels, a nonlinear calibration will be obtained after slit-scanning, and attention has been paid to linear regression methods by several authors (29-31). The conversion of the densitometer response is often more or less empirical. A generally valid regression method is hard to find, since the response depends on substance distribution and spot size. Since the calculations do not take much time, several regression methods may be tried for a particular chromatographic separation. The method with the best results can be chosen for subsequent chromatograms when the reproducibility of the procedure has been established. Signal conversion in this way is not restricted to HPTLC but may be performed in "classical" TLC as well, allowing linear responses for slit-scanning instruments.

In the case of larger chromatographic spots, it is more difficult to obtain reproducible results, and it may be necessary to leave simple one-wavelength measurements. The signal ratio at two (or more) wavelengths depends on the substance distribution, and this ratio implicitly offers a correction factor for the (too low) slit response. A dual-wavelength scanning device is incorporated in the Shimadzu CS 930. Actually, it is a pseudo-dual-wavelength device, whose principle is described by Ebel and Kussmaul (32) for comparing different measuring modes by computer simulation. The absorbance profile over the chromatographic track is measured twice at different wavelengths. At one wavelength the spots of interest show maximal absorbance, and at the other wavelegth the optical noise is measured. The signals are subtracted, resulting in effective reduction of optical noise. In future, (micro)computers will characterize developments in densitometry, especially in HPTLC.

Signal conversion and data handling for large numbers of samples will become more versatile and faster at lower prices, underlining the advantages of thin-layer chromatography. A very promising combination of a microcomputer and a densitometer was recently introduced by CAMAG (Muttenz, Switzerland). This topic is discussed in detail in Chapter 3 of this book.

ACKNOWLEDGMENT

The photoacoustic spectrum of silica was kindly recorded by Mr. F. J. van Werven, EG and G, The Netherlands. The author thanks Yvonne Louwers for enthusiastic help in typing the manuscript.

REFERENCES

1. E. Stahl. *Dünnschichtchromatographie,* Springer-Verlag, Berlin, 1967.
2. K. Macek. *Pharmaceutical Applications of Thin Layer Chromatography,* Elsevier, Amsterdam, 1972.
3. A. Zlatkis and R. E. Kaiser (eds.). *High Performance Thin Layer Chromatography,* Elsevier, Amsterdam, 1977.
4. W. Wendlandt and H. Hecht. *Reflectance Spectroscopy,* Interscience, New York, 1966.
5. J. B. Schute, H. J. de Jong, and H. A. Dingjan. Pharm. Weekbl. *105,* 1025 (1970).
6. H. Yamamoto. In: *Instrumental HPTLC*, R. E. Kaiser (ed.), p. 367, Hüttig Verlag, 1980.
7. S. Ebel. In: *Instrumental HPTLC,* R. E. Kaiser (ed.), p. 55, Hüttig Verlag, 1980.
8. P. Kubelka and F. Munk. Z. Tech. Phys. *12,* 593 (1931).
9. P. Kubelka. J. Opt. Soc. *38,* 448 (1948).
10. P. Kubelka. J. Opt. Soc. *44,* 330 (1954).
11. G. Kortüm. *Reflexions Spectroscopie,* Springer Verlag, Berlin, 1969.
12. F. A. Huf. Anal. Chim. Acta *90,* 143 (1977).
13. J. Goldman and R. Goodall. J. Chromatogr. *40,* 345 (1969).

14. R. W. Frei. Anal. Chim. Acta, *49*, 527 (1970).

15. A. Gleichen. *Lehrbuch der Geometrischen Optik*, Treubner, Leipzig, 1902.

16. F. A. Huf, H. J. de Jong, and J. B. Schute. Anal. Chim. Acta *85*, 341 (1976).

17. U. Hezel. Angew. Chem. *85*, 334 (1973).

18. S. Ebel and J. Hocke. Z. Anal. Chem. *277*, 105 (1975).

19. H. Jork. J. Chromatogr. *82*, 85 (1973).

20. L. R. Treiber. J. Chromatogr. *69*, 399 (1972).

21. F. A. Huf. Thesis, Leiden, 1978.

22. V. Pollak. In: *Densitometry in Thin Layer Chromatography*, J. C. Touchstone and J. Sherma (eds.), Wiley, New York, 1979.

23. *Shimadzu Instrument Manual.*

24. F. A. Huf. to be published.

25. L. R. Treiber, R. Nordberg, S. Lindstedt, and P. Stölln-berger. J Chromatogr. *63*, 211 (1971).

26. L. R. Treiber. J. Chromatogr. *123*, 23 (1976).

27. F. A. Huf, M. A. Span, and H. J. de Jong. In: *Advances in Steroid Analysis*, S. Göröged., Akadémiai Kiado, Budapest, 1982.

28. P. E. Jänchen. In: *Proceedings of the Second International Symposium on Instrumental HPTLC*, R. E. Kaiser (ed.), p. 220, Institute for Chromatography, Bad Dürkheim, Federal Republic of Germany, 1982.

29. R. E. Kaiser. In: *Proceedings of the Second International Symposium on Instrumental HPTLC*, R. E. Kaiser (ed.), p. 335, Institute for Chromatography, Bad Dürkheim, Federal Republic of Germany, 1982.

30. S. Ebel. In: *Proceedings of the Second International Symposium on Instrumental HPTLC*, R. E. Kaiser (ed.), p. 280, Institute for Chromatography, Bad Dürkheim, Federal Republic of Germany, 1982.

31. H. R. Schmutz. In: *Proceedings of the Second International Symposium on Instrumental HPTLC*, R. E. Kaiser (ed.), Institute for Chromatography, Bad Dürkheim, Federal Republic of Germany, 1982.

32. S. Ebel and H. Kussmaul. Fresenius Z. Anal. Chem. *269*, 10 (1974).

3

Computer-Assisted Quantitative Thin-Layer Chromatography and Its Applications in Screening Programs

MILICA KATIC, MIRKO PROSEK, and ARMIN MEDJA *Research Department, Lek Ljubljana, Ljubljana, Yugoslavia*

I. INTRODUCTION

The development of inexpensive microcomputers and the application of very good high-performance TLC (HPTLC) plates (1) have made new developments in quantitative TLC possible. We expect that the future of modern TLC lies in the intensive use of computers in all operations, especially in quantitative evaluation and in new analytical fields where its advantages can be shown.

We live in a time when people want to have as much information as possible for reliable decisions. This demand is really the advantage of TLC, because there is no other method that can give a comparable amount of information in such a short time, at the same low price as TLC (1). Usually this information is permanent because TLC plates can be rescanned several times. A plate contains a wealth of useful information, the utilization of which depends solely on our hardware and software. For example, if an HPTLC plate is developed for 5 min from both sides, with 20 samples each containing 10 components on either side, we can obtain 80 pieces of information per minute. This great output is the reason for on-line computer application in TLC. If we have a system which is able to scan a plate in a few minutes and calculate concentrations, we can acquire a complete evaluation of about 400 components in less than 10 min. The whole operation can be repeated if something goes wrong during the scanning. We see four potential fields of application: clinical chemistry, process control, bioengineering, and analysis of bioactive materials from plant tissues.

1. It is expected that the development of new assay procedures in clinical chemistry will help to optimize therapeutical doses by analyzing clinical samples such as plasma and urine in an inexpensive and simple way.

2. The modern computer-controlled process industry needs analytical methods that are quick, inexpensive, and not too sensitive to contaminated samples. In this application high-performance liquid chromatography (HPLC) has no advantage over TLC, as it is time-consuming, expensive, and too sensitive to contaminated samples. This field can be the domain of TLC. In this type of application the equipment producers must develop a totally automatic TLC system of an acceptable price. This will be difficult but not impossible.

3. The third field, bioengineering, is very promising. It is relatively young and is supported only by limited specialized analytical instrumentation. Initially, the development of analytical methodology is lagging behind, but this situation will change in the future. When the first euphoric jump is over, researchers will need suitable analytical capabilities, which may be TLC or some combination of TLC and electrophoresis. In these applications current HPLC is too slow and too delicate to be suitable for the analysis of thousands of crude samples.

4. The population of the world needs more and more food, and industrial processing of crops will continue to be important. Industrial production requires assay control, and TLC is very suitable for such a task.

These fields should benefit from TLC in the future because they all need methods that can give quick answers in cases where it is necessary to analyze a large number of samples in a short time, and where raw materials, without any complicated cleaning, must be evaluated.

Successful use of TLC in the fields described above is possible only if it is coupled to a computer. Without this, TLC will not be able to realize the promise it holds.

II. ON-LINE TLC

We started 10 years ago by connecting a Zeiss scanner to a desk top computer, model HP 9830A. The digitalized signal was sent to a computer through the BCD interface. This hardware allowed us to take two measurements in 1 s. With this system we examined and interpreted many chromatographic analyses. It was also used for theoretical research work, where we examined the propagation of light in thin layers (2-7).

This work gave us a knowledge of interfacing and programming. The development of low-priced microprocessors and microcomputers offered the possibility of using computer support in practically all fields of chromatography. A few years ago computers were much more expensive than analytical instruments. Today the opposite is true. It is possible to purchase a very sophisticated computer at a price that is only one-tenth of an instrument price.

The easiest way to use an on-line system is to buy such a system with complete hardware and software packages. In TLC it is difficult to find a suitable system, and chromatographers are forced to build their own.

A. Connection of a Computer to a TLC Scanner

In Figure 1 the connection between the computer and the scanner is shown. This connection can be very sophisticated. In such cases the computer controls all the functions of a densitometer. It is possible to have a simple connection where the computer is used as an analog recorder.

A sophisticated connection between a computer and a scanner can provide the following options:

1. Control over the movements of the scanning stage
2. Control of the monochromator
3. Selection of the lamp
4. Determination of the slit
5. Selection of the scanning mode (flying spot or slit)

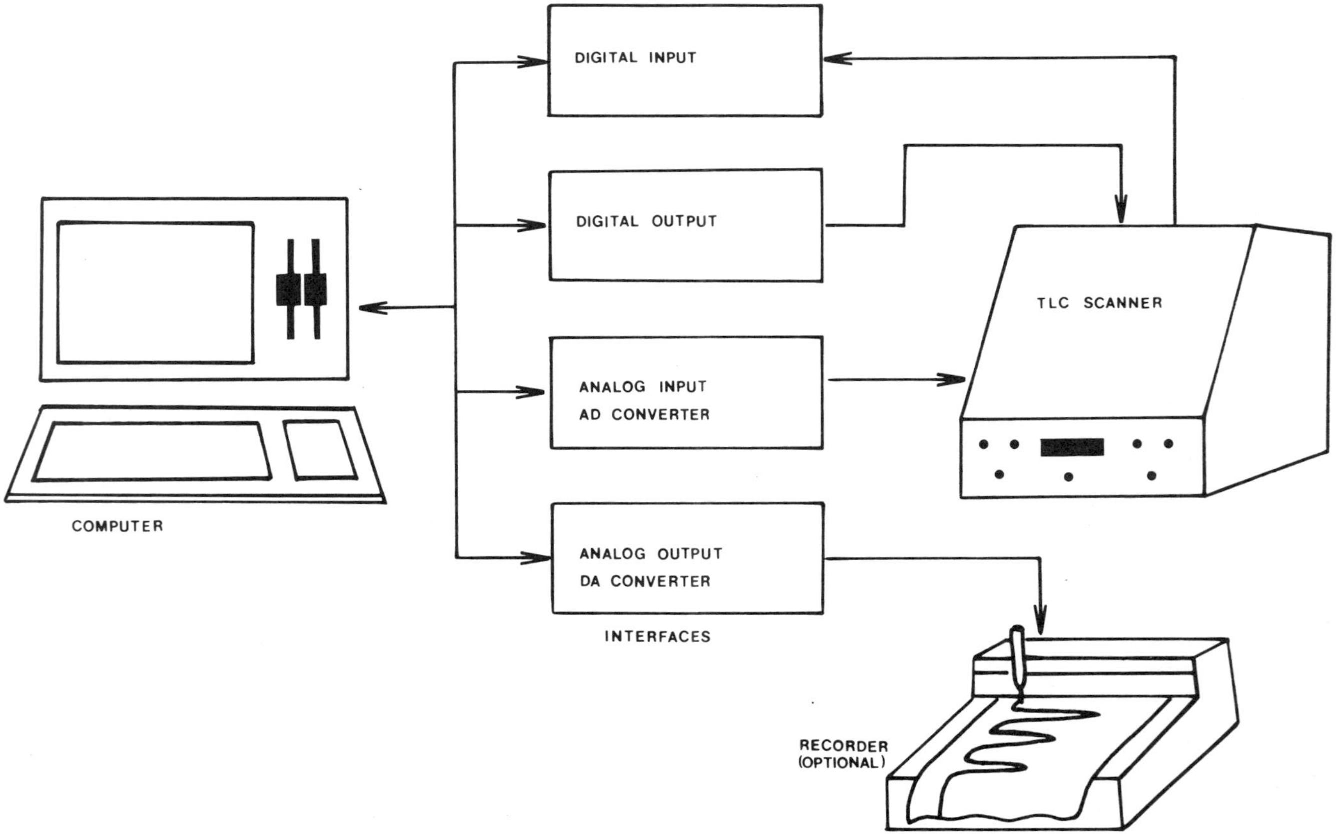

Figure 1 System for ON-LINE quantitative evaluation of TLC, with all necessary interfaces for sufficient control and evaluation.

6. Selection of gain
7. Selection of the scanning speed
8. Selection of the scanning length

When all these functions are implemented in hardware, we must prepare software which can perform a quantitative evaluation of TLC plates. Good software can communicate with the operator, usually a chemist, in a manner reasonably close to communication among chemists. It must control the scanner, take chromatograms, store raw data, integrate peaks, and perform postrun data manipulations, such as statistical calculations and the usual chromatographic calculations, including normalization based on internal and/or external standards.

In the next sections we describe some hardware and software used in on-line TLC.

III. HARDWARE

In this section the basic information about use of computers in the laboratory is discussed. A computer is a tool, and when the user is familiar with the tool the results of the work are much better.

The modes in which computers can be used in a laboratory may be classified in three groups: OFF-LINE, ON-LINE, and IN-LINE. OFF-LINE is the simplest mode, with no physical connection between the instrument and the computer. The data from the experiments are entered manually into the computer through a keyboard or any other device for data input on a standard stand-alone computer.

When the instrument is connected to a computer that controls the functions of the instrument and evaluates the measured values, but both devices are controlled by an operator, the mode is called ON-LINE. It gives the best results, but it needs trained people or a group of people who are familiar with the instrument and the computer.

The third version, the IN-LINE mode, is very common in modern instruments. In such an arrangement the computer is hidden from the operator inside the instrument, so that communication between the personnel and the machine takes place through buttons and knobs and not through the keyboard, as is usual with computers.

If we wish to construct ON-LINE quantitative evaluation of TLC we must connect a scanner (densitometer) to an adequate computer.

What are the characteristics of such computer? Is there any difference between the normal office computer and the laboratory computer? These two questions must be answered before we start the selection of a computer.

Basically, there is no difference between the computers used in the office and in the laboratory. A laboratory computer must be equipped with special components for communication between the computer and laboratory instruments. These components permit analog information (i.e., voltages, currents, or mechanical motions) to be translated into digital form suitable for the computer to store and manipulate. Digital-to-analog converters allow the reverse kind of translation, so that the digital information from the computer can be translated to analog values that can be sent to the experiment. It has input/output lines that can control on/off signals.

When one decides to buy a laboratory computer a system should be chosen that has built-in interfaces, or a system which can be equipped with additional interfaces at a reasonable price (8). Figure 2 shows a simple block diagram of a laboratory system. The most important part of a computer is the central processing unit (CPU), which represents the "brain" of the computer, because it controls the movement of data and instructions in the system. The programs are stored in the memory, which is available in two types: random access memory (RAM) and read only memory (ROM). Their names indicate the purpose of these memories. The ROM contains stored system programs, and this part of memory cannot be used for application programs. It is accessible to read from, but not to write into. For the latter purpose RAM is used. There are additional parts of the computer such as arithmetic and logic boards and special video processors, which help the CPU to be more successful in calculation and communication with external devices.

Each system consists of general input/output (I/O) devices and laboratory (process) input/output devices. The general devices are printers, terminals, disks (hard, floppy, winchester), plotters, digitalizers, tapes, etc. Special devices or interfaces are analog-to-digital converters (ADCs), digital-to-analog converters (DACs), binary I/O, timers, triggers, etc.

The special-purpose input/output devices, such as ADCs and DACs, are the most important parts of a laboratory computer system, and they are certainly a necessity if data acquisition and control of instruments are to be effectively implemented.

One very important aspect of using computers for laboratory experiments is that some, if not all, of the programming must be

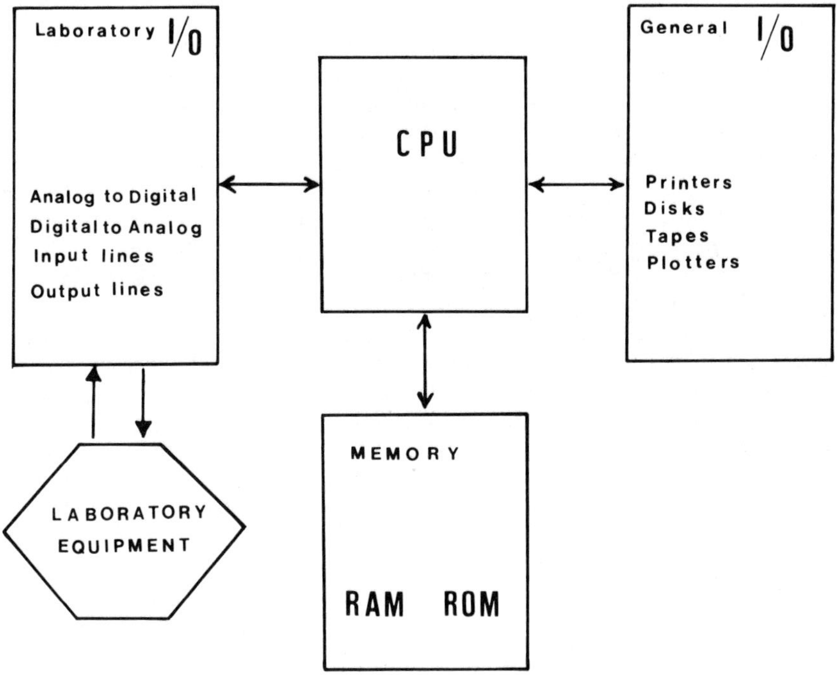

Figure 2 Block diagram of a typical laboratory system.

done in machine (or assembly) language, when a commercially available interface with adequate software is not selected. For some microcomputers (Apple II, IBM PC) interfaces with software programmable in BASIC are available.

The smallest information in a digital computer is a bit. It can have the values one (1) and zero (0). A group of bits forms a word; eight bits is called a byte. If a byte is a binary number, it may have a binary value of 00000000 to 11111111, which is equivalent to decimal 0 and 255. A great many modern computers use a byte or more bytes as a basic word of the system. In some computers there are different numbers of bits in a word. Computers with longer words are faster than those with shorter words, but not "wiser." Programs consist of a group of simple logic operations such as adding, subtracting, comparing, testing, and moving data through the CPU and the memory. The contents of a word do not indicate whether it is binary number or a coded representation of some symbol, operation, or instruction. Its meaning can be determined only by the context in which it exists. A skillful programmer can solve the same problem on a small eight-bit microcomputer as well as on big mainframe machines like an IBM, but the computation will take much longer on the microcomputer. In selecting a laboratory computer, one should know that any modern microcomputer can produce 200,000 operation in 1 s, which is usually enough for computer support in a laboratory. More computer power is needed only for making simulations of some process or searching for information from a large bank of data.

In TLC an eight-bit system can be selected, because it is powerful enough to solve all the problems that arise in this method and it is much cheaper than 16-bit microcomputers, especially the software. In the future 16-bit devices will be used in laboratories. For good resolution of analog signals at least 12 bits are needed. This means that two words of an 8-bit computer and only one word of a 16-bit computer are used for one measurement. As the combination of two words is inconvenient for programming, the 16-bit systems will find an opportunity in laboratories too.

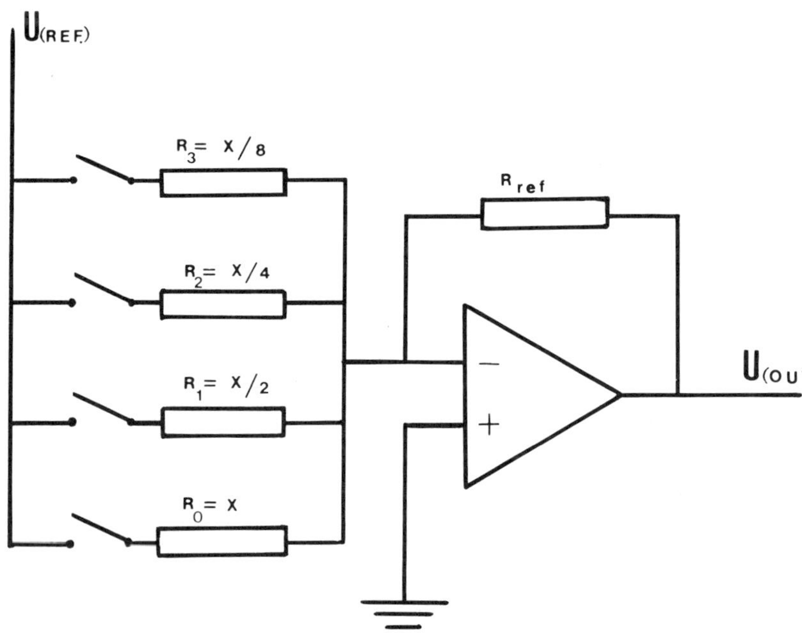

Figure 3 Diagram of a four-bit ADC constructed from a weighted resistor network.

A. Digital-to-Analog Converter

To be able to send an analog signal to an analog instrument from a computer, the computer must be equipped with a DAC. There are several different types of DACs. Figure 3 shows a diagram of a four-bit DAC constructed from a weighted resistor network, and from this picture it is possible to see the principle of all DACs. A digital word, which consists of zeros and ones, opens or closes the switches. The value of the current at the input of the operational amplifier is inversely proportional to the values of the connected resistors. U(out) depends on the sum of 1/R(x) and U(reference).

The analog value is obtained as follows:

$$U(out) = -U(ref)(D3/R3 + D2/R2 + D1/R1 + D0/R0)$$

The values of the resistor are selected in such a way that a linear range of analog values is obtained and each combination of bits can be transformed through the resistors and operating amplifier into an analog value. With a four-bit DAC it is possible to obtain 16 levels of analog output. When better resolution is required a DAC with a greater number of bits should be selected (see Table 1). The price of a DAC increases rapidly with the number of bits and with demands for linearity of the analog voltage, U(out).

B. Analog-to-Digital Converter

To convert an analog signal from an instrument, ADC is needed. Two types of converters are commonly used in the laboratory: the

Table 1 Resolution as a Function of Number of Bits

No. of bits	Resolution
2	4
4	16
8	256
10	1,025
12	4,096
14	16,384
16	65,536

successive approximation converter (SAR) and the dual-slope integrating converter. Both of them operate by comparing the measured voltage U(m) to a known reference voltage U(r); they differ principally in the mechanism used for generating U(r) and in the time required for a complete conversion.

The SAR is very fast and is very useful in applications where several thousand signals must be collected in 1 s, e.g., in mass spectrometry. For chromatography this is not the best type because the sampling time is very short, and when it is used for a sampling rate under 50 measurements per second a very noisy signal is obtained. For an acceptable signal, more measurements per data point must be taken and averaged.

The dual-slope integrating ADC is much more suitable for such applications. Its operation is completely different from that of the SAR. The diagram of such an ADC is shown in Figure 4, which compares two voltages and two times: U(m), the measured value; U(r), the reference value; T(on), the time it takes to charge a capacitor C with the measured voltage U(m) (this time is fixed and is adjustable in some converters); and T(off) the time required to discharge the capacitor when it is connected to the reference voltage U(r) of opposite polarity.

When the converter is started, U(m) is connected through R to the capacitor C, which begins to charge. This operation lasts for a determined period of time (counts) T(on). After that U(m) is disconnected and U(r), which has opposite polarity, is connected to the capacitor C. Now the counter counts the time needed for C to be fully discharged.

The time required for total discharge is proportional to U(m). Since the time of charging is constant, the following equation applies:

$$U(m)*T(on) = T(off)*U(r)$$

For a given ADC, T(on) and U(r) are constants and can be selected to give T(off) in known units (e.g., volts). This value is used further by the computer.

This type of ADC measures relatively long intervals, some milliseconds, and it acts as an average of the voltage. Positive spikes charge the capacitor more rapidly and the negative spikes more slowly, so that at the end of one measurement the capacitor C or T(off) presents a medium value of U(m).

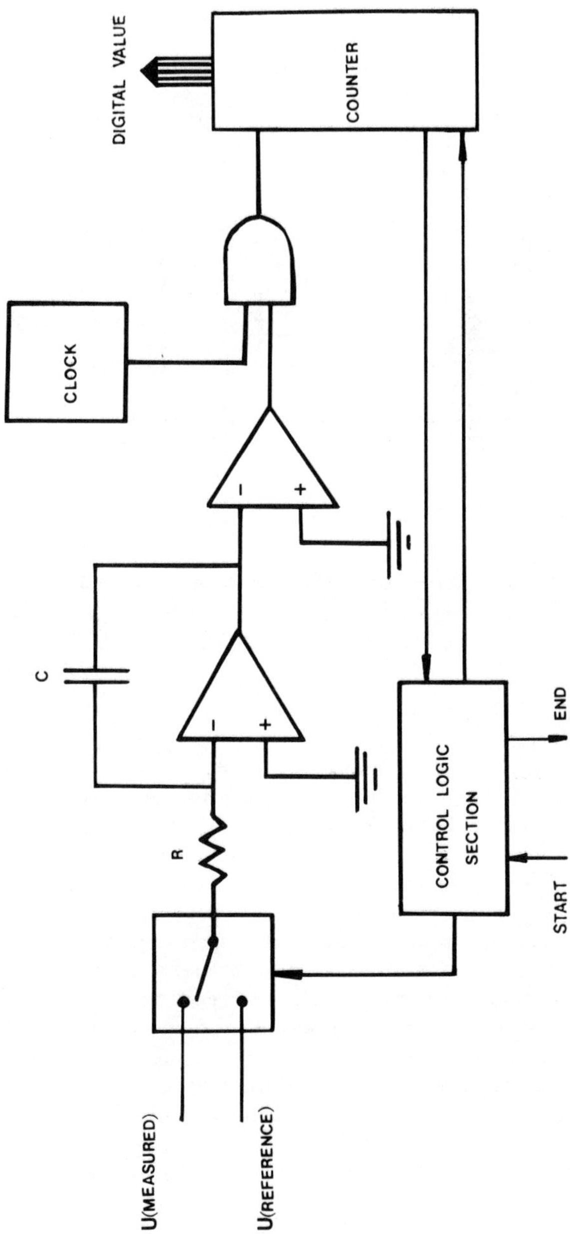

Figure 4 Diagram of a dual-slope integrating ADC. In this type of converter the time necessary to discharge a capacitor C is T(off).

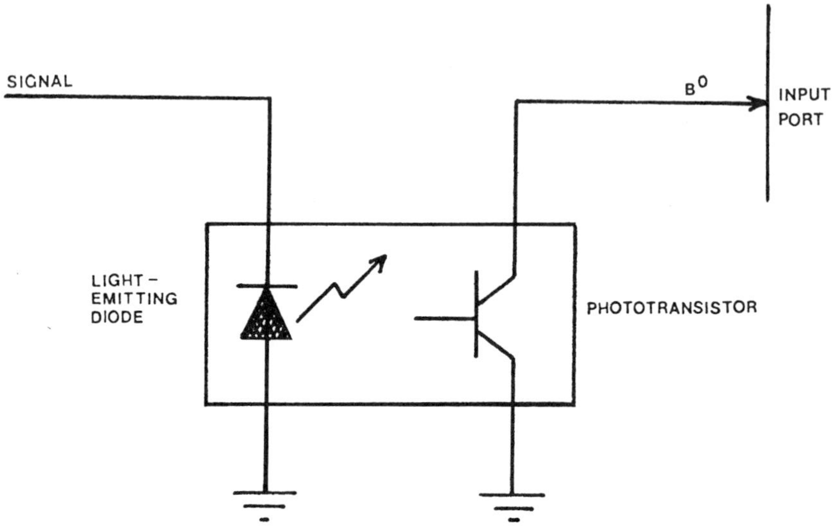

Figure 5 Input line BO is separated from output signals with an optocoupler. A light-emitting diode activates the optotransistor in the case of active signals.

C. Digital I/O

Input/output lines are very important parts of a laboratory computer. Through those lines the computer is able to control external devices and to obtain information about logic changes in them.

Input lines are separated from the computer through the optocouplers (Figure 5) in which IC inputs are electrically isolated from outputs. On one side of the line (input) there is a light-emitting diode (LED). On the other side there is a photosensitive diode, which responds to the light beam from the LED. A change in the logic state of the input lines changes the emission of the LED, and the light beam transfers the information from an external device into the computer. The protection of the input lines from external voltage variations is very good, usually more than 2000 volts.

The output lines are also protected from external voltage variations, but the problem of these lines is the outputs, which must be amplified from low-level computer signals. For this purpose normal relays with amplifiers, triacs, or solid-state relays are used. These lines must be shielded. When a high voltage is turned on or off near the computer, electromagnetic interferences (EMI) usually occur. Such interference can cause computer to malfunction. Together with the hardware, a very important part of the computer is the software for I/O applications, because these lines are used for synchronization of the programs in the computer with external events.

Usually, laboratory computers are equipped with an internal crystal clock, which is used as a clock, a calendar, and a time base in the experiments.

IV. SOFTWARE

In laboratory applications the software plays a very important role. It is expensive because it is written in machine code and is highly specialized. In this section we describe the basic ideas of programs used for the evaluation of data in chromatography, particularly in TLC.

The software for chromatography can be divided into two groups. The first group is older, and includes programs that use some sort of data reduction during the analysis in order to run on a simple computer configuration. These programs are not very useful in TLC because they detect peaks, compute the area under curves, and perform area allocation for overlapping peaks at almost the same time that the data are being acquired from the densitometer. As this program cannot correctly predict the influence of the baseline on the peaks, the evaluation of the results is usually not correct, especially in TLC, where the influence of the plate structure on peak areas is significant. The programs in the second group evaluate the chromatograms after the data have been collected. These programs are much more useful for TLC.

Quantitative determinations in chromatography, including TLC, depend on correct evaluation of peak heights and areas under the peaks. It is essential that the beginning and the end of the peaks are accurately detected. Such detection is difficult since, in spite of the electronic filtering, the signal usually contains noise, the effect of which is particularly sensitive in the region of the baseline. Baseline drift, if appreciable due to the structure of the plate, further complicates the problem of detecting the beginning of a peak, especially when the peak is wide and of low amplitude.

Numerous programs and publications have been devoted to the problem of poorly separated peaks. In practical work it is always easier and better to improve the chromatographic separation of incompletely resolved peaks rather than manipulate them by computer methods, however clever they may be.

Even with the most sophisticated analog filtering, some low-frequency noise and spikes reach the computer. It is essential, therefore, that some form of digital smoothing be applied to the incoming data before the real-time analysis programs are employed. Two types of digital smoothing programs are commonly used: the least-squares fit and group averaging. In our programs we use both methods, but we have found that simple averaging of 2, 4, or 8 points gives us better results than more complicated methods using polynomials. As the averaging can be performed in a much shorter time than a least-squares fit, and the computing is critical element in small systems, we usually use averaging.

Various methods are employed for detecting peaks, but the vast majority of them are based on the use of first or second derivatives. All methods work reliably with large peaks with sharp leading edges. Difficulties are experienced with smaller or distorted peaks. Methods employing first derivatives are less noise-prone than those employing derivatives. Therefore, for the detection of small peaks in an equivalent signal-to-noise environment, the first derivative method is better than the second. On the other hand, use of the second derivative gives the rate of change of slope and so is independent of baseline drift.

The first derivative peak detection criterion recognizes a peak when the first derivative rises above a set threshold. To avoid the deteciton of spurious peaks it is used to take more successively computed values of the first derivative to be greater than the set threshold before a peak is said to have been detected. The detection logic of a second derivative requires at least three sample points over the threshold (five measurements) to be positive. In TLC, it has been observed that peaks whose shapes are seriously distorted from the Gaussian peak shape cannot be detected by routines employing the second derivative.

For quantitative TLC we developed our own version of the well-known routine for peak start identification (9), and it works very well. It detects peaks by looking for an increasing angle from the single sample point on the baseline. This routine has the advantage of being independent of baseline drift and peak shape (Figure 6).

The return to the baseline after a peak has been detected and analyzed (peak maximum, Rt) also has a significant effect on the accuracy of the area measurement. Again, both the first and second derivative methods can be used. A return to the baseline is detected when the second derivative is zero. As this baseline

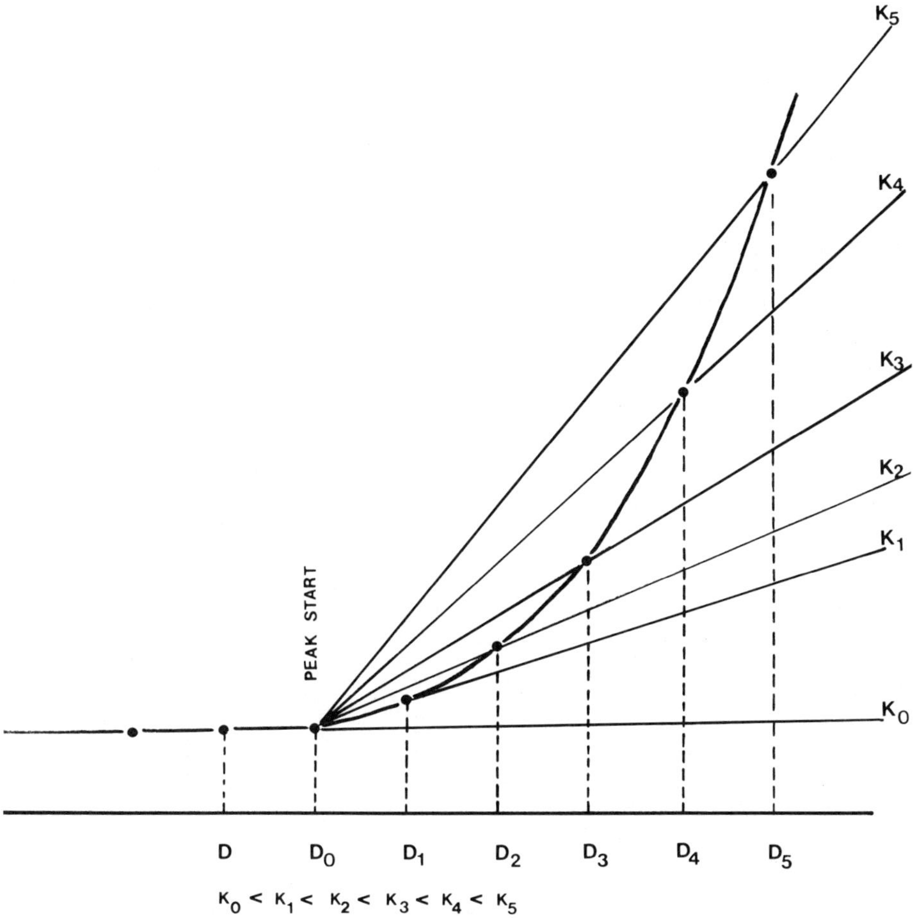

Figure 6 Routine for peak detection. The peak is detected when the increasing angle from the signal sample point appears five times.

detection criterion can be fulfilled at erroneous points, some additional restrictions must be used. Very often, these restrictions include the height of the signal. When the signal is higher than the expected baseline, the return to the baseline cannot be detected. In a real-time integration mode, when peaks are evaluated during the data acquisition, baseline recognition and allocation of peak shapes within a peak train are a very difficult task, which

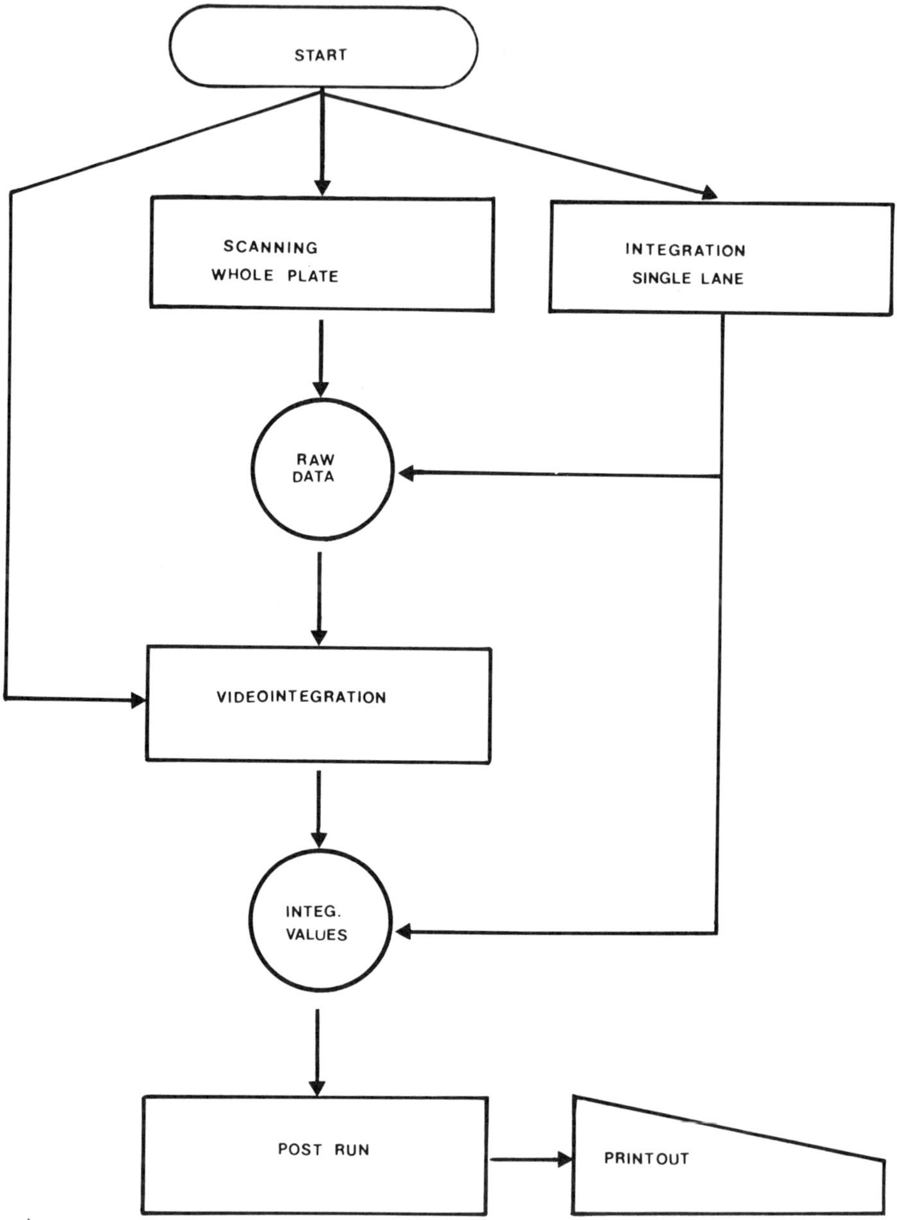

Figure 7 IfC-LEK TLC pack.

can often be performed incorrectly. In a postrun integration, when peaks are evaluated after the run and from the whole scan, this operation is much easier and more valid.

There are situations where, without doubt, the most efficient detector is the human eye. For such cases we prepared a group of programs that offer the advantage of postrun integration but have additional possibilities. Special commands allowing the operator to influence the evaluation and allocation of peak areas have been added.

In our laboratory a package consisting of about 20 programs to be used in different applications with different types of scanners has been compiled. It is called the IfC-LEK TLC pack. The basic system (Figure 7) is divided into three levels: (a) scanning of raw data, (b) determination of chromatographic parameters and integration, and (c) postrun data handling. At the first level, there are programs that control the scanners, take care of data acquisition, and store the raw data on a disk. These programs are written partly in high-level language and partly in machine language. All the subprograms for ADC and I/O routines of scanners are written in machine language. This concept is flexible because it is only necessary to change these subroutines in order to modify the program for a new type of scanner or a new type of interface or ADC. All the programs look the same to the operator, who has no access to the subprograms.

The IfC-LEK TLC pack was prepared for on-line TLC and HPTLC analysis on an APPLE II plus microcomputer with 48K, equipped with an Adalab universal interface or some other (12–16 bit) ADC. The pack was produced at the Institute for Chromatography Bad Dürkheim, B.R.D., and the Research Department of the Pharmaceutical Factory LEK Ljubljana, Yugoslavia (10).

An outstanding feature of this software system is the simplicity with which it can be installed and applied to the task of processing TLC data. It can be used in a batch mode, with the whole TLC plate taken as a unit, or in a single-lane mode, with a lane taken as a unit. It includes some programs that simulate the flying-spot scanning mode on different types of densitometers (11).

The TLC pack is designed to operate in two modes, manual and automatic. In the manual mode the data acquisition is started manually on the computer and on the scanner separately, and the end of data acquisition is determined by the computer. In the automatic mode a special program controls the scanner.

The only program that is able to scan and integrate in one run is the IFC integrator. It uses subprobrams from levels one and two, and changes an Apple II with an additional 48K into a single-channel integrator. To perform this function the analog densitometer signal is digitalized. The raw data are stored in the Apple II memory. The noise is eliminated with a digital filter. The data for evaluation of the chromatogram are computed. The program performs peak detection, area measurement, and retention time measurements, and computes peak area percentage. After the integrated values are printed out on a printer, they may be stored together with the raw data (chromatogram) on a disk for further processing by other programs. The IfC integrator is a very simple device designed for routine work with chromatograms that are not very complicated.

At level two we have a group of programs which are used for postrun integration of raw data that have been collected by a data acquisition program from level one. These programs give the operator complete control over the chromatograms. Communciation between the data and the operator is provided by the cursor on the CRT. It is possible to construct a baseline to enter peak starts and peak ends. Some versions have a built-in algorithm for automatic determination of peak parameters. When the operator is satisfied with the peak parameters the program proceeds to the integration. The integrated areas are displayed on the CRT, so it is possible to see the baseline and peak shapes. Automatically determined peaks are displayed in white and manually determined peaks in gray. The calculated values are shown and printed out in the form of the analytical report. The data are saved on a disk for further calculation with POST RUN DATA HANDLING.

At level three the final chromatographic calculations are performed. This program offers four calculation procedures, which can be performed on integrated values obtained from IfC INTEGRATOR AND VIDEOINTEGRATOR. These caluclations are:

Area %
Normalization
External standard
Internal standard

The communication between the program and the operator is clearly described in the information line, and the operator can work with the program.

With adequate selection of programs a whole system can be constructed which is capable of collecting data, integrating, and printing the results. At present, it is possible to select between Adalab, Tecmar ADC, and Mountain ADC + DAC. This pack can be used on Zeiss PMQ2 and PMQ3, Camag, Kontes, and Sigma 20 scanners. scanners.

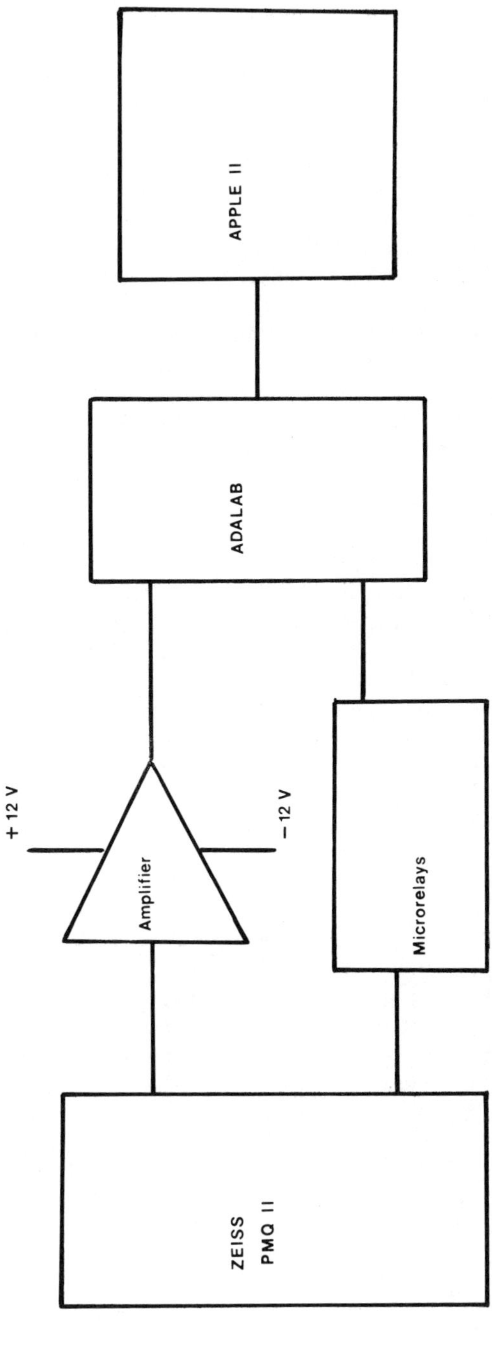

Figure 8 Connection of Zeiss PMQ2 densitometer and Apple II microcomputer.

In our factory, quantitative evaluations are performed with two different types of on-line systems. Figure 8 shows the Zeiss PMQ densitometer connected to an Apple II computer via a universal Adalab interface board (Interactive Microware Inc., P.O. Box 771, State College, PA 16801). This system is designed specifically for laboratory applications and has one A/D, one D/A, and 16 I/O digital lines. Practically any instrument able to send or receive analog signals can be connected to the Adalab board.

The low-output signals of the densitometer (0 to 50 mV) require amplification before they can be fed into the ADC. The construction of a special board incorporating microrelays for control purposes ensured complete separation of the densitometer from the computer and thus prevented the instrument from affecting computer operation. As the PMQ2 was not originally equipped for external control, all the switches had to be fitted with electric relays. Manual search of lane position was selected because it was much faster than the automatic mode, since the influence of noncentral location of a spot in the slit gave rise to errors and necessitated many very small, time-consuming steps in automatic search routines. This decision was not a drawback for use of the system, as it should always be remembered that the computer is an analytical aid and should not become the master of the analytical method. The automation should not be regarded as an end in itself.

Since the Kontes densitometer has precise control in the Y direction, we added a step motor to the X direction and obtained control over two-dimensional movement. Figure 9 shows the Kontes densitometer equipped with an additional step motor in the X direction. It is controlled by an Apple II microcomputer equipped with a Tecmar DA101 digital-to-analog converter (Tecmar Inc., 23600 Mercatile Road, Cleveland, OH 44122) and a Tecmar AD211 analog-to-digital converter. The analog signal is digitalized by the Tecmar AD211, an interface designed for sophistical laboratory or industrial applications. In the standard version it has either 16 single-ended or 8 true differential inputs. The resolution and accuracy is 12 bits. It has a conversion rate of 30 kHz. The Tecmar DA101 is interfaced with two independent high-speed digital-to-analog converters. In addition, it provides eight parallel output lines, which can be used for the external control of the recorder and densitometer. In this on-line combination, it is possible to control scanning speed and length of the scan in both directions by the number of pulses from the interface. Two relays were wired parallel to the toggles on the front panel in order to gain control over the scanning head position and the direction of the scan.

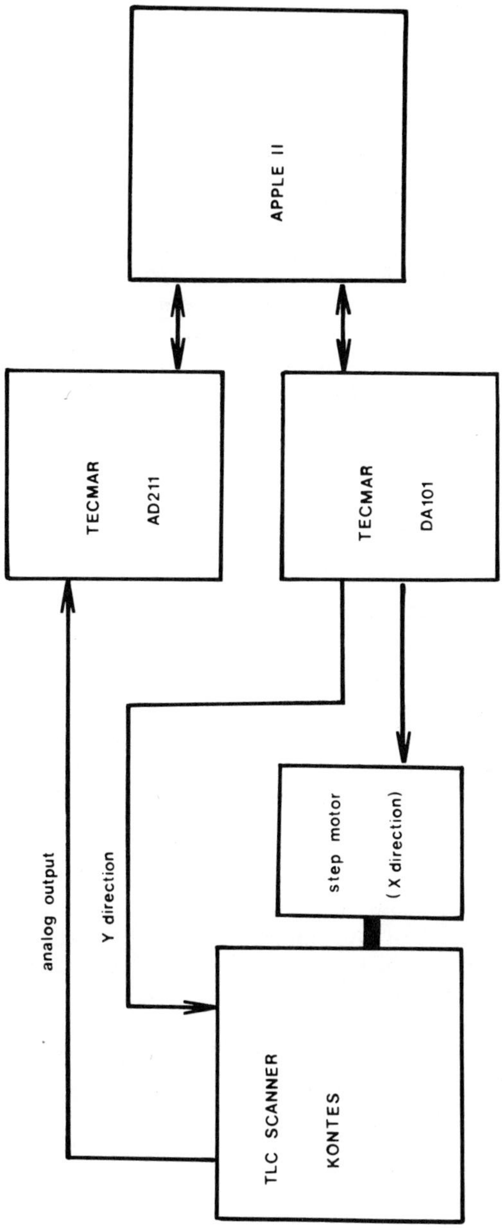

Figure 9 Connection of Kontes densitometer and Apple II microcomputer.

The two systems described above have been used in quantitative evaluation of many plates prepared in our laboratory. They were used in the determination of degradation products and by-products of hydrogenated ergot alkaloids (12), and were also useful in the determination of active substances in extracts from plants.

In addition, the great potential of on-line quantitative HPTLC as a sensitive precise micromethod can be seen in the determination of metoclopramide in plasma, which was demonstrated in our Research Department. We developed two rapid methods for the analysis of metoclopramide in human plasma. Metoclopramide was isolated by extraction with ethyl acetate at an elevated pH. The extracts from human plasma were applied on commercially available silica gel plates. Quantitative results were obtained with diazotization and VIS measurement on a Kontes scanner. This method allows the detection of 10-ng metoclopramide in 1 ml of plasma with a standard deviation of ±5% (13).

About 100 times better results were obtained with HPTLC plates and fluorescence measurement on a Zeiss PMQ scanner. In this mode less than 20 pg of standard has been detected and about 50 pg of active substance in plasma extract was evaluated without any interference of the baseline.

The most powerful of all quantitations performed with on-line TLC are the screening methods. They can very successfully replace the visual inspection usually used in such procedures, because they are faster, more objective, and much more sensitive than usual methods.

In a recent publication we described the on-line quantitative TLC evaluation of large amounts of thebaine in *Papaver bracteatum* (14). This work was performed in order to enrich the content of thebaine in poppy. This is a difficult task because it is necessary to examine a large number of plants in a relatively short time, between the collecticn of the poppy and the preparation of new fields. In order to solve this problem within the time constraint, on-line TLC was selected.

A set of programs was written that was subsequently combined into a document called POPPY pack. The programs were written in Applesoft (Microsoft Basic). All I/O routines were prepared in machine code with a Merlin macroassembler. These subroutines were executed by CALL statements. After testing, the programs were compiled. Since our programs contain many graphic routines and integer calculations, the compilation was very successful, and it increased the execution speed by a factor of 10. The Microsoft compiler TASC provides an important extension of Applesoft, the

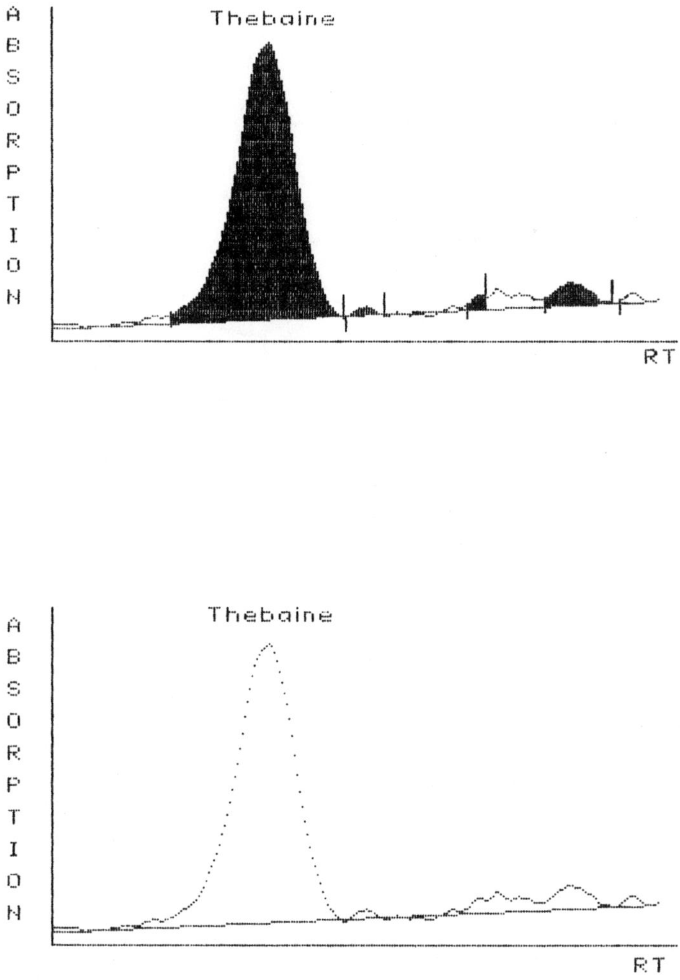

Figure 10 Densitogram of an extract of *Papaver bracteatum*.

use of COMMON variables. COMMON is implemented by allocating a static block of memory for a selected number of data. This block of memory is protected, and all the programs in our pack use the same data. Once the data are collected, they can be processed by any compiled program which uses the COMMON variables. A whole block of data can be transferred to or from the disk into an

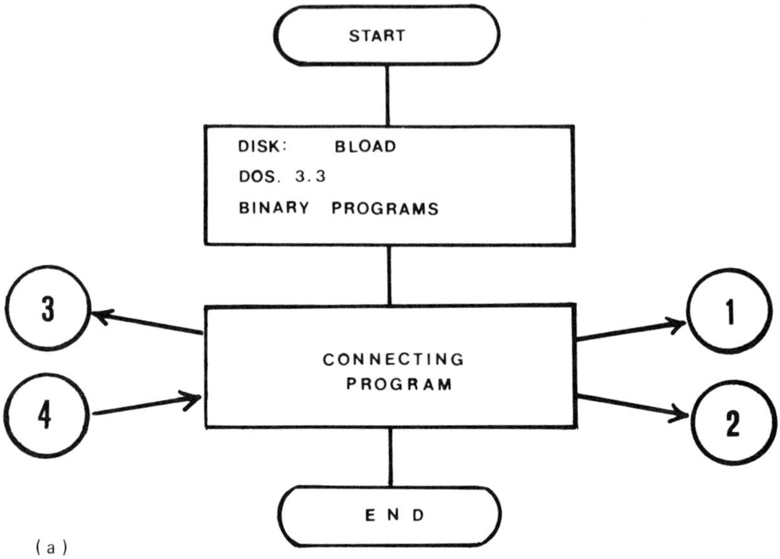

Figure 11 Flowchart of programs used in quantitative evaluation of tebaine.

allocated area in binary form, which also increases the speed of data processing.

The start of the program is "Hello." It is used to combine the whole pack, and it stores all binary programs for I/O routines and the Library program of the compiler. Additional tasks in the pack are selected through the menu. The operator can select among:

1. Scanning
2. Integration
3. Data handling

When two disk drives are used, the system disk remains in drive 1 and the data disk is inserted in drive 2. In a one-drive system the disk must be replaced by a data disk. The data disk is used only for storage of raw data and integrated values. The final results are stored on the system disk, because these data are used in additional statistical and printout programs.

In the scanning program a whole plate containing up to 40 samples is taken as one run. The names of the individual lanes on a plate are entered and stored in the text file. The measured data are stored in binary form. A number of measurements for one

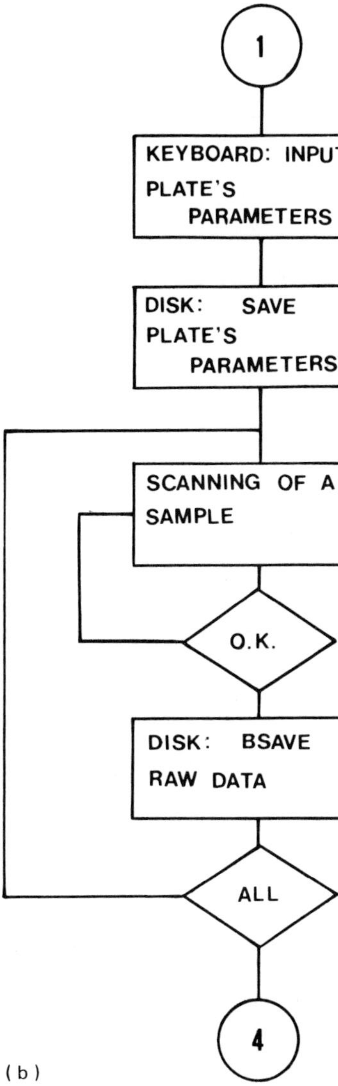

(b)

Figure 11 (Continued).

data point can be selected from 1 to 10. With this option the scanning distance and influence of noise are controlled. Scanning was performed at 280 nm, and 190 data points were collected. With the

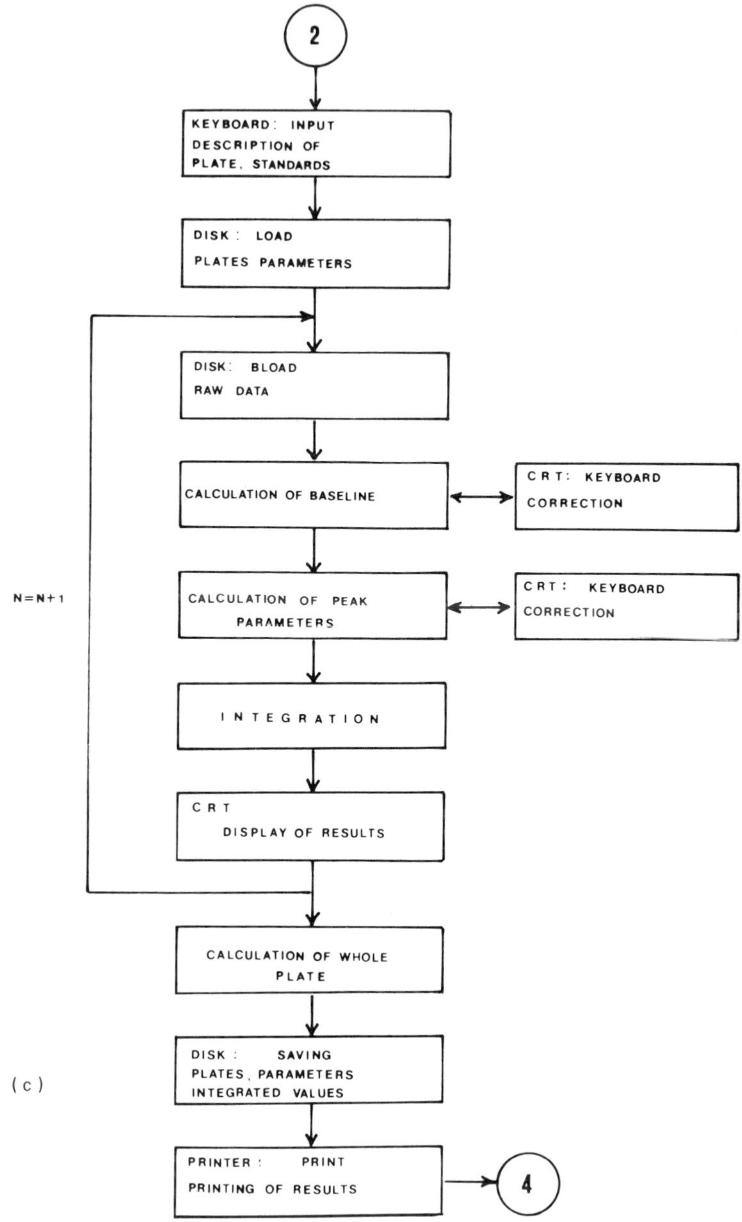

Figure 11 (Continued)

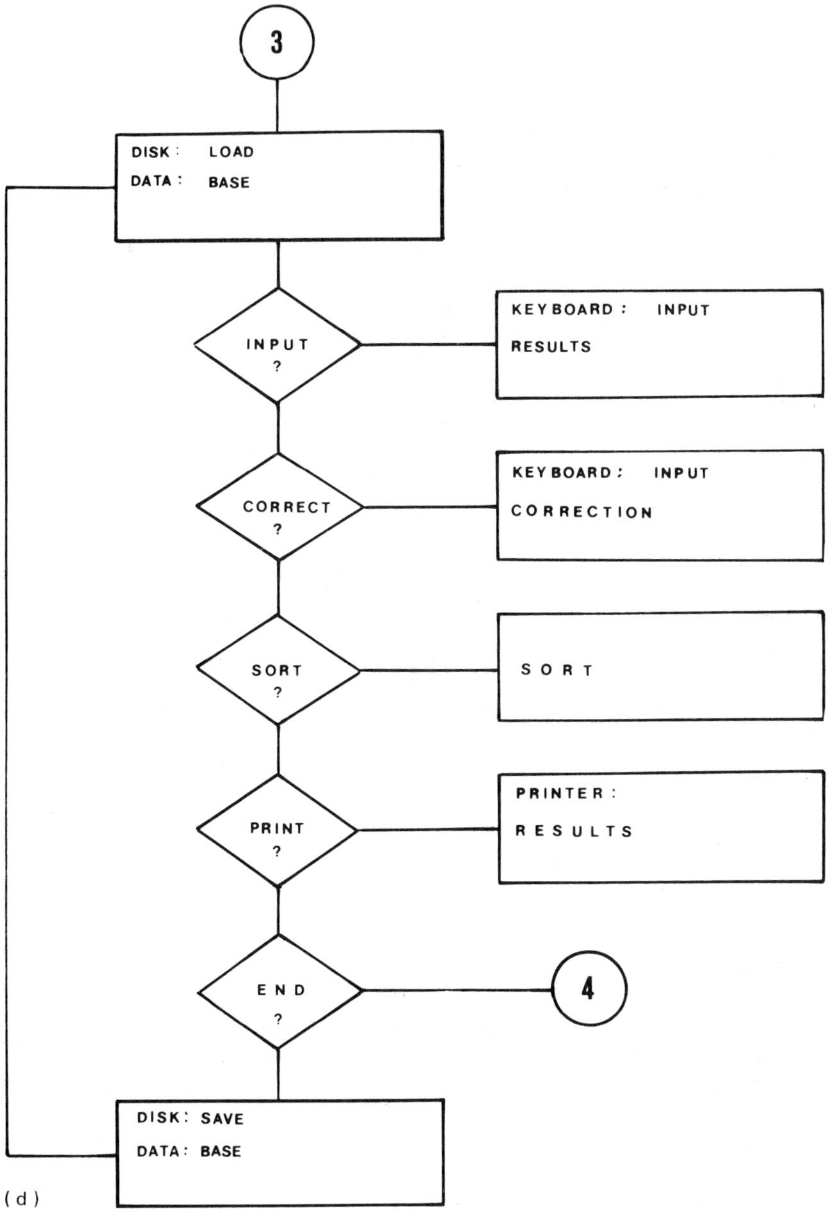

Figure 11 (Continued)

good extraction and efficient chromatography system used, this reduction of data was acceptable (Figure 10).

The algorithm for integration is the same as that in our TLC pack, but some simplifications were added to permit the high speed of determination. The most important features of this program are baseline correction and isolation of the correct peak. Very good separations allowed us to use the largest peak in the lane as the sample instead of searching for it according to R_f values. In rare cases where there was a higher peak than the one examined, the former was eliminated by a baseline routine; it was declared a baseline. Usually these samples had such small amounts of active component, that they were not important. The "dialog" between the operator and the measured values was performed through graphics and instructions on the screen. After the last lane on the plate was processed, standards were detected and a calibration curve was calculated. In the next step each sample was evaluated and printed out in form of the analytical report. Our first target was the construction of a closed system with raw data as the input and the final selection as the output, but later we changed this approach. The final results for the samples were added to the data base from the keyboard. This was a better solution, because in cases having low levels of the active substance in the sample, the result was added to the data base without additional testing. In cases having commercially interesting concentrations, other and more sophisticated analytical procedures were undertaken, the results of which were entered into the data base.

Postrun data handling allows different approaches to the data base. It was possible to sort data according to parameters such as concentration, location of the farm, and time of production. Reports were printed out in form of a book.

Figure 11 shows flowcharts of these programs.

V. CONCLUSION

On-line evaluation of TLC data with a low-priced personal computer equipped with an inexpensive I/O interface can help us to solve many problems. We were able to scan five plates (half-plate) with 17 lanes each corresponding to a total of 85 chromatograms, in 1 hr. The postrun evaluation took an additional 5 min. In cases with a good baseline about 100 to 120 samples, adding up to 300 to 400 densitograms, can be evaluated.

On-line processing has additional advantages. The result can be printed out and stored in files. Our group spent many hours in creating this pack, but when the speed and accuracy of the selections performed in our laboratory are taken into account, the results were well worth our effort.

For the future we expect new generations of low-priced super personal computers in the same category as the AMICA and ATARI 1040 ST and Apple MAC Plus. These generations of computers will represent new capabilities that are highly desirable for everyday laboratory work and process control. The development of new types of scanners with diode array sensors or multiarray phototransistors, together with new microcomputers combined with sophisticated software, will open up new possibilities for real communication between the operator and the samples. For these reasons, chromatographers should be familiar with microcomputing. From now on we have opportunities, but we must accept the challenge, and we must transfer the knowledge of many generations of TLC chromatographers into the programs.

REFERENCES

1. A. Zlatkis and R. E. Kaiser. *High Performance Thin Layer Chromatography*, Elsevier, Amsterdam, 1977.
2. M. Prosek, A. Medja, E. Kucan, M. Katic, and M. Bano. HRC & CC *2*, 517 (1979).
3. M. Prosek, A. Medja, E. Kucan, M. Katic, and M. Bano. HRC & CC *2*, 661 (1979).
4. M. Prosek, A. Medja, E. Kucan, M. Katic, and M. Bano. HRC & CC *3*, 183 (1980).
5. M. Prosek, A. Medja, E. Kucan, M. Katic, and M. Bano. HRC & CC *4*, 138 (1981).
6. M. Prosek, A. Medja, E. Kucan, M. Katic, and M. Bano. HRC & CC *4*, 141 (1981).
7. S. Lenarcic, M. Prosek, A. Medja, E. Kucan, M. Katic, and M. Bano. HRC & CC *5*, 694 (1982).
8. J. J. Carr, *Microprocessor Interfacing*, TAB Books, 1982.
9. E. F. G. Woerlee and J. C. Mol. J. Chromatogr. Sci. *18*, 258 (1980).
10. R. E. Kaiser and M. Prosek. LaborPraxis *1982*, 310.

11. M. Prosek, M. Katic, M. Kucan, A. Medja, and M. Bano. Anal. Chem. Symp. Ser. *13*, 27 (1983).

12. M. Prosek, M. Katic, and I. Verbic. CAL *3*, 223 (1983).

13. M. Katic, M. Prosek, A. Medja, A. Anzur, and E. Kucan. CAL *1*, 50 (1983).

4

Quantitative Thin-Layer Chromatography with Flame Ionization Detection

KUMAR D. MUKHERJEE *Federal Center for Lipid Research, Münster, Federal Republic of Germany*

I. INTRODUCTION

Thin-layer chromatography (TLC) is probably the simplest and most economical chromatographic technique for fast separation of mixtures of substances (1), yet quantitation of the components resolved by TLC is not always accomplished satisfactorily (2, 3). Scanning of thin-layer chromatograms in situ by spectrophotometric techniques is considered the method of choice for routine analysis of substances that exhibit fluorescence or absorb ultraviolet (UV)/visible light (4, 5). These in situ spectrophotometric techniques are, however, of limited use in the quantitative TLC of substances that contain only weak or no chromophoric groups, such as lipids and many other aliphatic and alicyclic compounds (3). For such substances, the procedures that have been proposed for routine analysis by TLC include staining of the resolved spots with suitable reagents followed by in situ quantitation by spectrophotometric techniques (6). In all these techniques, extensive empirical calibration is a prerequisite for adequate accuracy. Thus, in photodensitometry of carbonized spots of organic substances, separated by TLC, it is mandatory to chromatograph several known mixtures along with different amounts of sample to be analyzed (7, 8).

Some of the alternative techniques used for quantitative TLC of organic substances that contain weak or no chromophoric groups involve their elution followed by (a) gravimetric determination (9, 10), (b) conversion to chromophoric derivatives and subsequent spectrophotometric determination (11–13), and (c) conversion to volatile derivatives and subsequent gas chromatographic determination (14–16). All these techniques are fairly accurate, but rather tedious for the routine analysis of large numbers of samples.

Since the late 1960s, attention has been focused on the in situ detection and quantitation of substances, resolved by TLC, with the aid of vapor phase detectors that are commonly used in gas chromatography (2, 17). Three types of techniques for quantitative TLC with vapor phase detectors have become known so far.

In one set of techniques, the substances resolved on the adsorbent layer are vaporized and swept by a carrier gas to a vapor phase detector, such as a flame ionization detector (FID) or thermal conductivity detector (TCD). In an early device working on this principle (18, 19), TLC is carried out on narrow quartz plates coated with silica gel. After removal of the developing solvent, the chromatoplate is encased in a rectangular quartz chamber and driven through a pyrolysis furnace, while a carrier gas, nitrogen, flows through the chamber to an FID. Thereby, the substances separated on the chromatoplate are consecutively vaporized by pyrolysis and/or evaporation and the gaseous products are swept by the carrier gas to the FID, which records each fraction as a separate peak.

The above technique was considerably improved by replacing the chromatoplate, encased in the quartz chamber, by a *chromatotube*, i.e., a quartz tube whose internal wall is coated with a thin layer of silica gel or some other inorganic adsorbent (20). The mixture to be analyzed is separated in such chromatotubes in a manner similar to the separation on chromatoplates, after which the developing solvent is removed by heating in a stream of nitrogen. The chromatotubes are then scanned by driving them through a tubular pyrolytic furnace (800°C), while a carrier gas, nitrogen, flows through the chromatotubes into an FID. During scanning, the individual fractions resolved in the chromatotube are consecutively vaporized, swept by the carrier gas to the FID, and monitored. The technique of TLC with chromatotubes (20), later termed *tubular TLC* (TTLC) (2, 17), is a versatile method for separation and quantitation of less volatile and weakly chromophoric substances that are difficult to analyze either by direct gas chromatography or by TLC and in situ spectrophotometry.

In a later modification of tubular TLC (21, 22), the mixtures to be analyzed are separated in chromatotubes made of glass, whose internal wall is coated with an adsorbent such as silica gel and an oxidant such as cupric oxide. After removal of the developing solvent, the chromatotubes are scanned by driving them through a tubular combustion furnace (650°C), while a carrier gas, helium, flows through the chromatotube. Thereby, the organic substances, fractionated in the chromatotube, are consecutively combusted in situ to carbon dioxide and water. The gaseous combustion products and any uncombusted vaporized organic material are swept by the carrier gas through a heated (650°C) tube, filled with cupric oxide, in which complete combustion of all organic material occurs. Thereafter, the carrier gas stream flows through a water trap into a TCD, which monitors the carbon dioxide evolved from the individual fractions in the chromatotube as separate peaks whose areas correspond to the carbon content of each resolved fraction.

Techniques and instrumentation to be described later will demonstrate how the principles of pyrolysis and combustion can be combined for quantitation of organic substances separated by tubular TLC.

As opposed to the above systems, in which the substances separated by TLC are vaporized from the adsorbent layer and transported by a carrier gas to a vapor phase detector, another set of techniques have become known in which TLC is carried out on an adsorbent-coated quartz rods (23) or metal strips (24). The resulting *Chromatorods* or *Chromatostrips* are then scanned, in order to detect and quantitate the separated components, by driving them directly through the flame jet of an FID, which records the individual fractions as separate peaks. These techniques are reminiscent of the systems using FID in conjunction with a moving chain (25) or a moving wire (26) for the detection of eluates from liquid chromatography.

In a third group of techniques (27, 28) for quantitative TLC with vapor phase detectors, separation of mixtures is carried out on an adsorbent-coated disk, and the eluate emerging from the adsorbent is continuously conveyed by a moving wire to an FID and monitored in a manner similar to that in liquid chromatography (26).

After this overview of the development of various vapor phase detection systems for quantitative TLC, the main part of this chapter will be devoted to a detailed description and critical evaluation of the techniques of tubular TLC and coated-rod TLC in conjunction with flame ionization detection and their present as well as potential industrial applications.

II. TUBULAR TLC-FID SYSTEMS

The general techniques used in these systems involve separation of mixtures on internally coated chromatotubes and subsequent detection of the resolved fractions by driving the chromatotubes through a furnace, whereby the components are consecutively vaporized and transported by a carrier gas to an FID.

A. Preparation of Chromatotubes (29)

Quartz tubes or high-melting glass (Pyrex) tubes, 20 cm long, having a 6-mm outer diameter and 5-mm inner diameter, are generally used. The tubes are cleaned by immersing them in a bath of chromic acid or a laboratory detergent and thoroughly rinsing with tap water followed by deionized water.

For scanning by pyrolysis, the chromatotubes are generally prepared as follows. Silica gel G (E. Merck, Darmstadt, Federal Republic of Germany) is slurried with water at a ratio of 1:2 (wt/vol) for about 2 min. The quartz tube is held vertically with one end dipped into the slurry. Suction is applied from the other end to draw the slurry almost to the top, and then suction is released to allow the slurry to drain. This leaves a uniform layer of adsorbent on the internal wall of the tube. The coated tubes are then held vertically for about 5 min until the adsorbent layer becomes firm. Thereafter, the chromatotubes are activated by heating in an oven at 110–130°C for 1–2 h.

For scanning by combustion, the chromatotubes are generally prepared as follows. Cupric oxide used as an oxidant is usually preheated at 650°C for 2 h, cooled, ground, and sieved to a size of 200/270 mesh. Silica gel G containing 15–20% cupric oxide is commonly used for coating the glass tubes. A slurry of the adsorbent containing the oxidant is prepared, either in water at a ratio of 1:2 (wt/vol) or in a mixture of chloroform, methanol, and water (50:50:2, by vol) at a ratio of 1:1.9 (wt/vol), and the glass tubes are coated as described above. The chromatotubes are finally activated by heating at 110–130°C for 1–2 h.

Before using the chromatotubes for separations, any organic matter present on the adsorbent layer should be removed. This is accomplished conveniently in the tubular TLC scanner, described below, by prescanning the chromatotubes at a high speed (3–4 cm/min) with the tubular furnace kept at 800°C (for quartz chromatotubes) or 650°C (for glass chromatotubes). The prescanned chromatotubes can be stored in a desiccator for prolonged periods.

B. Chromatography (17, 29)

With a microsyringe, a few microliters of the solution containing up to 50 μg of the mixture to be analyzed are spotted onto the adsorbent layer at a distance of about 1–2 cm from one end of the chromatotube. After evaporation of the solvent, the chromatotube is developed in a chamber, such as a screw-capped (Teflon-lined) test tube or a glass-stoppered measuring cylinder containing a few milliliters of the developing solvent. After the solvent front has reached the desired height, the chromatotube is removed from the chamber. The developing solvent is quantitavely removed from the adsorbent layer by blowing a gentle stream of nitrogen through the chromatotube, which is held at an elevated temperature (100–120°C) in an oven. The chromatotube is then ready for scanning.

C. Construction and Operation of the Tubular TLC Scanner

The simplified scheme (17) in Figure 1 illustrates the working principle of the tubular TLC scanner (Packard Model 7251, TLC

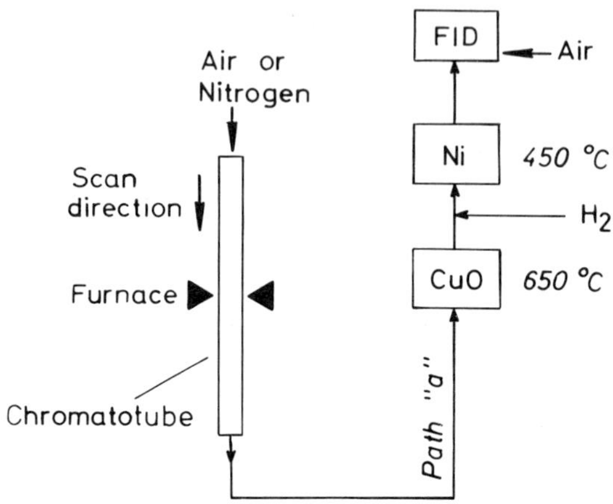

Figure 1 Working principle of the tubular TLC scanner. (Reproduced in part from Ref. 32 with permission of Elsevier Publishing Co., Amsterdam, The Netherlands.)

QUANTUSCAN, built by Packard-Becker, Delft, The Netherlands; inquiries concerning this instrument should be addressed to Fischer Labor und Verfahrenstechnik, Heerstrasse 35-37, D-5300 Bonn-Bad Godesberg, Federal Republic of Germany. The scanner has been designed in such a manner that in situ quantitation of substances separated on the chromatotubes can be performed by both pyrolysis (on quartz tubes coated with silica gel) and combustion (on glass tubes coated with silica gel and cupric oxide) (29, 30). The essential features of the tubular TLC scanner are the driving unit and the detection unit.

The construction of the driving unit is shown schematically in Figure 2 (30). The chromatotube is mounted vertically in the driving frame with its upper end connected to the inlet of the carrier gas, generally nitrogen, and with its lower end connected to the detection system via compressible capillary spring or narrow Teflon tubing. The upper and lower joints of the chromatotube are sealed gastight with gaskets made of a heat-resistant septum of the type commonly used in gas chromatography. All the gas lines connecting the chromatotube with the detection system are kept heated (200–250°C). With the aid of a variable-speed motor the chromatotube is driven at a desired scan speed (0.5--5

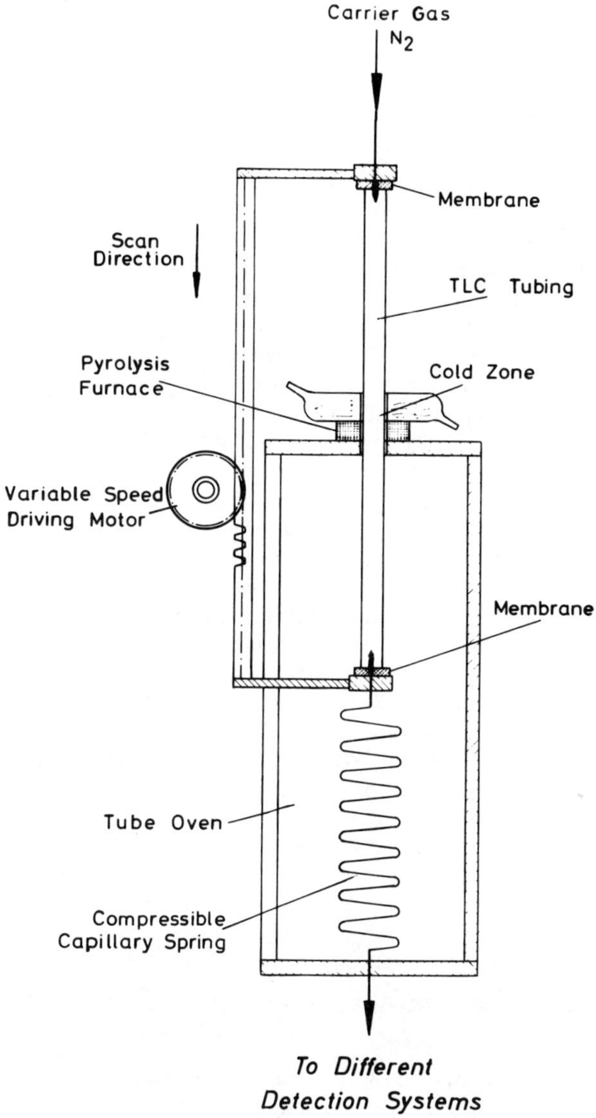

Figure 2 Driving unit of the tubular TLC scanner. (Reproduced from Ref. 30 with permission of Elsevier Publishing Co., Amsterdam, The Netherlands.)

cm/min) through the narrow (2 mm) cylindrical furnace into a heated (200–250°C) tube oven. The latter prevents condensation or readsorption on the adsorbent layer of any organic material vaporized from the chromatotube.

While the chromatotube traverses the scanning furnace, the organic substances separated on the adsorbent layer are consecutively vaporized and swept by the carrier gas (40–45 ml/min) to the detection system. A cylindrical cooler made of quartz can be mounted on top of the scanning furnace in order to achieve a steep temperature gradient on the chromatotube in the vicinity of the heating zone. The scanning furnace is held at 800°C when the organic substances are vaporized from the adsorbent by pyrolysis, or at 650°C when they are converted to gaseous carbon dioxide and water by combustion on the adsorbent containing cupric oxide. Some evaporation of the substances from the adsorbent layer may occur during scanning by either pyrolysis or combustion.

The carrier gas containing the vaporized products form the chromatotube is fed continuously to the detection system, as shown schematically in Figure 1. First, the carrier gas enters a heated (650°C) oxidation tube, packed with cupric oxide, where all the organic matter vaporized from the chromatotube is quantitatively combusted to carbon dioxide and water. Thereafter, a stream of hydrogen (40–45 ml/min) is mixed with the carrier gas, and the resulting gas stream enters a heated (400–500°C) reduction tube packed with Chromosorb-Raney Nickel catalyst (31). In the reduction tube carbon dioxide is reduced to methane, which is subsequently quantitated in the FID. A strip chart recorder connected with the FID amplifier registers the FID signals, which appear as separate peaks for the individual fractions resolved in the chromatotube. The peak areas correspond to the proportions of methane or carbon dioxide derived from each chromatographic fraction, which, in turn, correspond to the total carbon content of each fraction.

D. Practical Considerations in Quantitation by Tubular TLC

Successful application of tubular TLC in the quantitative analysis of mixtures of organic substances is governed by the proper choice of the chromatographic system and consideration of the factors that affect sensitivity and reproducibility. These aspects are considered in the following discussion, and a few typical examples of practical applications of tubular TLC are given.

1. *Characteristics of the Chromatotubes (29)*

The signal recorded by the FID of a tubular TLC scanner is due to all the combustible organic substances present in the chromatotube. Thus, not only the components of a mixture fractionated in the chromatotube but also any organic contaminant eventually present in the adsorbent would give rise to an FID signal. For a reliable quantitation, the chromatotubes should be devoid of foreign organic matter, which might cause undesirably high background signals of the FID. Apart from prescanning the chromatotubes prior to analysis, utmost care should be taken to avoid organic contamination during sample application and chromatography.

Separation patterns in chromatotubes coated with silica gel alone or silica gel containing 15–20% cupric oxide are essentially identical to those observed on conventional chromatoplates coated with silica gel. Most of the solvent systems commonly employed for various separations by conventional TLC can be used with minor modifications, if required, for TLC on chromatotubes.

Reusability of chromatotubes is an important factor governing the application of tubular TLC in routine work. Chromatotubes coated with silica gel alone can be used repeatedly for at least 10 times, because after each scanning the adsorbent layer is virtually devoid of any organic matter and its activity is barely altered. The chromatotubes coated with silica gel and cupric oxide, however, cannot be reused outright, because after scanning the cupric oxide present in the areas of the separated fractions is reduced to metallic copper. For reuse, such chromatotubes can be rescanned at a high speed (2–4 cm/min) with a furnace kept at 650°C and air or oxygen streaming through them; thus the areas of reduced copper are reoxidized. Such treatments barely affect the activity of the adsorbent layer, and the chromatotubes can be used repeatedly for at least 10 times.

If air is used as the carrier gas for scanning chromatotubes, cupric oxide, which is reduced to metallic copper due to in situ combustion, is reoxidized in the presence of excess oxygen (32). Such chromatotubes can be reused outright without any further treatment.

2. *Factors Affecting Quantitative Results (29)*

The sample size should be chosen such that no overloading of the chromatotube occurs. Moreover, if the amount of substance in a fraction exceeds a certain limit, the FID signal is no longer linear with the amount of substance and errors in quantitation may result. For lipids and related substances, the amount of substance per chromatographic fraction should not generally exceed 10–20 μg.

The lower limit of detection and quantitation of organic substances separated in chromatotubes is governed not only by the efficiency of the systems for combustion to carbon dioxide and subsequent reduction to methane and by the sensitivity of the FID, but also by the organic contaminants on the adsorbent layer that cause the background FID signal. With low-background chromatotubes as little as 0.1 μg of substance in each chromatographic fraction can be detected and quantitated with fairly good accuracy. Scan speeds generally chosen are 1–2 cm/min, which ensure complete vaporization of most high molecular weight compounds from the chromatotube by pyrolysis at 800°C or by combustion at 650°C. Higher scan speeds can be applied, which might necessitate raising the temperature of the scanning furnace for complete vaporization of certain substances. Although high scan speeds shorten the time of analysis, they may cause overlapping of peaks of adjacent fractions and overloading of the FID with consequent nonlinearity of the FID signal.

Performance of the conversion systems of the tubular TLC scanner, i.e., efficiency of combustion in the oxidation tube and of reduction of carbon dioxide to methane in the reduction tube, should be checked from time to time. This can be done conveniently by injecting known volumes of carbon dioxide or methane into the conversion system and recording the FID signal. Oxidation efficiency is generally over 99.9%. The efficiency of reduction of carbon dioxide to methane generally ranges from 70 to 95%, depending on the activity of the Raney Nickel catalyst, temperature, and flow rate of the carrier gas. The reduction efficiency is reproducible under a given set of conditions (standard deviation ±0.1%).

3. *Methods of Quantitation and General Applications*

A great merit of tubular TLC in quantitative analysis lies in the fact that the peak areas recorded by the FID correspond to the absolute carbon content of the organic substances located in each fraction of the chromatotube (29, 30). Therefore, in general, the response factors of the individual components of a mixture separated by tubular TLC need not be determined by empirical calibration with mixtures of known composition, as in many other techniques of quantitative TLC. Instead, the response factor of each compound can simply be calculated from its elemental formula.

Table 1 Quantitative Analysis of a Mixture of Lipids by Tubular TLC

Compound	Weight % (known)	Composition determined through scanning by[a,b] Pyrolysis	Combustion
Cholesterol	54.0	55.1 ± 0.8	55.8 ± 1.7
Tripalmitoylglycerol	46.0	44.9 ± 0.8	44.2 ± 1.7

[a]Chromatographic conditions: adsorbent for pyrolysis, Silica Gel G; adsorbent for combustion, Silica Gel G containing 15% cupric oxide; developing solvent, hexane/diethyl ether (80:20, by vol); sample size, 10 μg. Quantitation conditions: scan speed, 1 cm/min; temperatures: scanning furnace, 800°C (pyrolysis), 650°C (combustion); tube oven and lines, 250°C; oxidation tube, 650°C; reduction tube, 500°C; flow rates: carrier gas (N_2), 45 ml/min; hydrogen, 45 ml/min; air, 700 ml/min.
[b]Values are means ± S.D.
Source: Reproduced, in part, from Ref. 29 with permission of the Journal of Chromatographic Science, Niles, Ill.

Table 1 shows an example of quantitation of a mixture of cholesterol and tripalmitoylglycerol analyzed by tubular TLC, using the techniques of pyrolysis in situ as well as combustion in situ (30). In these analyses the response factors are calculated as follows. Since the combustion of each component is essentially complete, 1 μg each of cholesterol and tripalmitoylglycerol yield stoichiometrically 3.07 and 2.77 μg, respectively, of carbon dioxide, which is proportionally reduced to methane and detected by the FID. Hence, with respect to cholesterol as unit, the response factor for tripalmitoylglycerol amounts to 3.07/2.77 = 1.108. When the observed peak areas of cholesterol and tripalmitoylglycerol are corrected by multiplying with the response factors of 1.000 and 1.108, respectively, the composition of the mixture can be calculated. The data given in Table 1 show that the values determined by tubular TLC are in close agreement with the actual composition of the mixture.

If natural mixtures are resolved by TLC, a particular chromatographic fraction representing one class of substances may be composed of several homologous compounds. If tubular TLC is applied

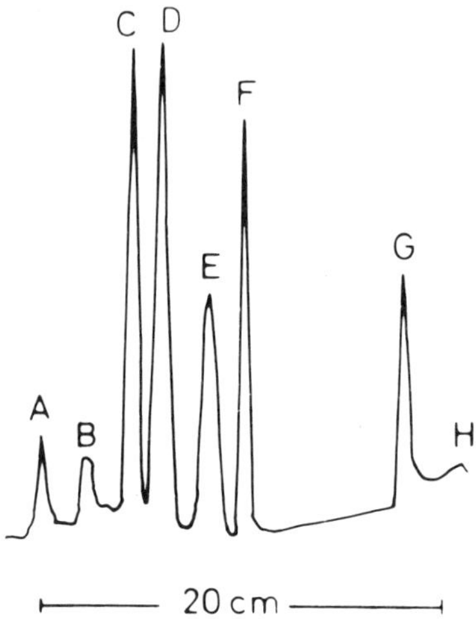

Figure 3 Chromatogram of human skin surface lipids fractionated by tubular TLC and scanned by pyrolysis. Peak identification: A, polar lipids; B, sterols; C, unesterified fatty acids; D, triacylglycerols; E, stearyl esters plus wax esters; F, squalene; G, paraffins; H, solvent front. Chromatographic conditions: adsorbent, silica Gel G (E. Merck); developing solvents: first development with hexane/diethyl ether/formic acid (80:20:1, by vol) up to 10 cm; second development with hexane up to 20 cm; sample size, 40 μg. Quantitation conditions: scan speed, 1 cm/min; temperatures: scanning furnace, 800°C; tube oven and lines, 250°C; oxidation tube, 650°C; reduction tube, 550°C; flow rates: carrier gas (N_2), 45 ml/min; hydrogen, 45 ml/min; air, 700 ml/min. (Reproduced from Ref. 29 with permission of the Journal of Chromatographic Science, Niles, Ill.)

for the analysis of such mixtures, the average yield of carbon dioxide from each class of homologous compounds should be known in order to calculate the response factors accurately (29, 30).

Tubular TLC has been applied in the quantitative analysis of a wide variety of lipid mixtures (20–22, 29, 30, 33–35). A few typical scans of lipid mixtures fractionated and quantitated by tubular TLC are shown in Figures 3 and 4. Some industrial applications of tubular TLC will be discussed later.

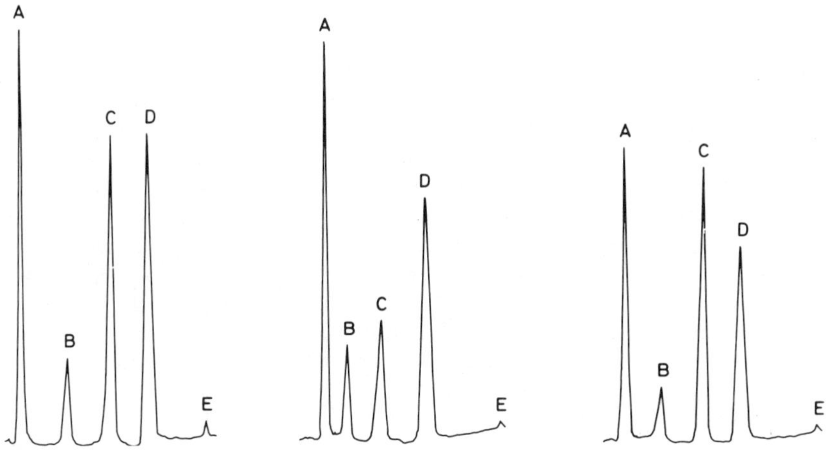

Figure 4 Chromatograms of different samples of human serum lipids fractionated by tubular TLC and scanned by combustion. Peak identification: A, polar lipids; B, cholesterol; C, triacylglycerols plus unesterified fatty acids; D, cholesteryl esters. Chromatographic conditions: adsorbent, Silica Gel G containing 20% cupric oxide; developing solvent, hexane/diethyl ether (85:15, by vol); sample size, 20 μg. Quantitation conditions: identical to conditions given in Fig. 3 except for the temperature of the scanning furnace, which was 650°C. (Reproduced from Ref. 29 with permission of the Journal of Chromatographic Science, Niles, Ill.)

III. COATED-ROD TLC-FID SYSTEMS

The original technique involving separation of mixtures on adsorbent-coated glass rods and subsequent detection of the separated components by driving the chromatorod through the flame jet of an FID (23) has been substantially improved since the development of reusable quartz rods that are coated with the adsorbent in porous sintered glass (36, 37). The application of this technique in the analysis of lipids has been reviewed (38).

A. Preparation of Coated Rods

Quartz rods coated with an adsorbent in sintered glass are generally prepared as follows (36). One part of silica gel or alumina and three parts of fine glass powder are suspended in a solvent, such as acetone or ethanol, and a thin quartz rod of the desired length is dipped into the suspension and gently withdrawn. The quartz rod is thus uniformly coated with a thin layer of the mixture of adsorbent and glass powder. The rod is then air-dried and baked in a furnace at 800–1000°C, whereby a rigidly adhering layer of the adsorbent in porous sintered glass is obtained.

Three types of coated quartz rods carrying layers of adsorbent in sintered glass are commercially available under the name Chromarod (worldwide distributors: Newman-Howells Associates Ltd., Wolvesey Palace, Winchester, Hants, S023 9NB England). These are 0.9-mm-diameter quartz rods having a length of 152 mm, of which the entire length, except 2 mm at each end, is coated with a 75-μm layer of special frit of soft glass and adsorbent. Chromarod-S and Chromarod-SII are coated with silica gel with particle sizes of about 10 and 5 μm, respectively; Chromarod-A is coated with alumina with a particle size of about 10 μm.

B. Chromatography

The Chromarod is mounted on a rod holder that accommodates up to 10 rods. The same rod holder can be used for developing the Chromarods and subsequent scanning. The mixture to be analyzed is dissolved in a suitable solvent and spotted near one end of the Chromarod by using a microsyringe or micropipette. Generally 5–10 μg of the sample is adequate for analysis. After evaporation of the solvent, the rod holder carrying the Chromarods is placed in a common TLC chamber containing the solvent, and the Chromarods are developed up to the desired height. The rod holder is then taken out of the chamber and the solvent is removed from the adsorbent layer by heating. The Chromarod is then ready for scanning.

C. Construction and Operation of the Coated-Rod TLC Scanner

The construction of the coated-rod TLC scanner, Iatroscan TH-10 MK III series (worldwide distributors: Newman-Howells Associates Ltd., Wolvesey Palace, Winchester, Hants, SO23 9NB England), is shown schematically in Figure 5. The rod holder carrying the developed Chromarods is mounted in the scanner, in which each Chromarod is consecutively driven at a chosen scan speed (0.13–0.73 cm/s) from one end to the other through the flame jet of the FID. During scanning, the components resolved on each Chroma-

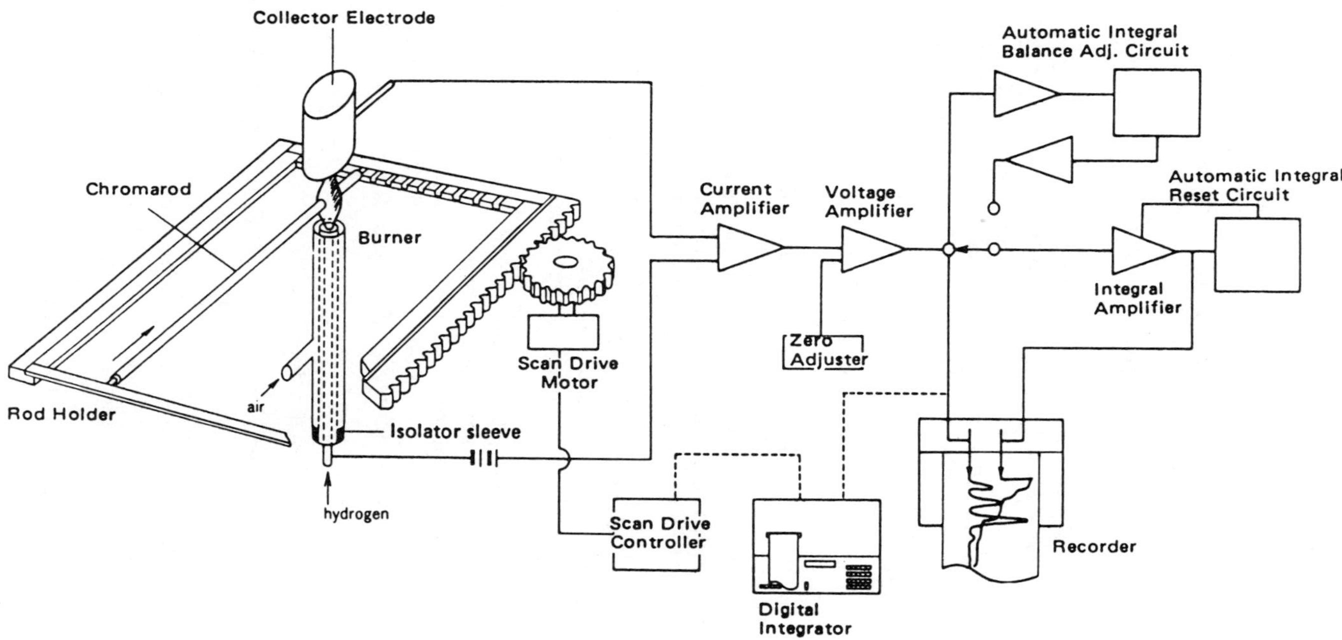

Figure 5 Construction of a coated-rod TLC scanner, Iatroscan TH-10 MKIII series. (Reproduced with permission of Newman-Howells Associates Ltd.)

rod are vaporized/pyrolyzed one after the other and the ionizable carbon is converted to ions, which are detected in collector electrode. The signals generated by the collector electrode are passed through to an amplifier, integrator, and recorder.

D. Practical Considerations in Quantitation by Coated-Rod TLC

Several factors deserve careful consideration before coated-rod TLC can be used satisfactorily for the quantitative analysis of mixtures of organic substances. The following discussion includes the characteristics of coated rods, factors affecting quantitative results, as well as methods of quantitation and general applications.

1. *Characteristics of the Coated Rods*

As in tubular TLC, all precautions must be taken in coated-rod TLC to prevent contamination of the adsorbent layer with organic impurities that yield a high FID background signal, which severely hampers quantitation. It has been recommended (38) that Chromarods to be used for the first time should be "cleaned" by passing them twice through the flame jet of the FID in the scanner. Subsequently, the Chromarods should be stored in a humidified chamber free from organic matter. Prior to use, the Chromarods should be run once again through the flame jet of the FID. Used Chromarods should be periodically (after 10 to 15 analyses) cleaned in chromic acid and washed in running tap water and finally in distilled water.

For a given adsorbent, the order of migration of substances of varying polarity on Chromarods broadly resembles their migration pattern on corresponding TLC plates. Therefore, most of the solvent systems used for a particular separation by conventional TLC can be applied with minor modificaitons for separations on Chromarods.

2. *Factors Affecting Quantitative Results*

Simplicity, speed, and commercial availability of precoated rods are some of the attributes that prompted many researchers to explore the techniques of coated-rod TLC in the analysis of various mixtures. Especially in the field of lipids, there have been many studies on the separation of complex mixtures on coated rods and numerous recorder tracings of such separations have been presented (39–46, and the relevant literature cited in Ref. 38). Yet it was not until recently that the reliability and reproducibility of quantitative results obtained by coated-rod TLC-FID techniques were critically evaluated and the factors affecting quantitation were studied in some depth (47–54).

As opposed to tubular TLC-FID techniques, where the response factors for the individual components are governed only by their carbon contents (29), in coated-rod TLC-FID techniques considerable variations in the response factor for the same substance are frequently observed (47, 48, 52, 55, 56). It should be emphasized that the response of the FID for the components separated on coated rods is not always governed by the same physicochemical phenomena as in gas chromatography. In gas chromatography, the components quantitated by the FID are always present in a vaporized state, and in general only the "ionizable carbon" contained in the molecule gives rise to the FID signal (57–59). In the coated-rod TLC-FID system, however, the components resolved are volatilized in the flame jet partly by physical evaporation and partly by pyrolysis, which leads to breakdown of the parent molecule. Consequently, the FID response may not correlate with the ionizable carbon theoretically present in the parent molecule. Other factors that affect the FID response of substances resolved on coated rods include sample size, scan speed, and rod uniformity.

Response factors have been found to increase, often nonlinearly, with either increasing (48, 52) or decreasing (53) sample weight. Moreover, the composition of a mixture has been found to affect the response factors of the individual components (47, 52).

Variability of response factors due to nonuniformity of commercially available precoated rods is a phenomenon frequently observed by users of coated-rod TLC-FID techniques (38, 48, 52, 54). Impregnation of precoated rods with 5% copper(II) sulfate solution prior to analysis has been found to minimize rod-to-rod variation and improve baseline stability (54).

Among the operating variables of the coated-rod TLC scanner, the flow rate of hydrogen for the FID should be optimized, to begin with, to ensure a stable baseline of the FID signal and good reproducibility of the response factors (53, 56). Scan speed is a crucial operating variable which can greatly affect the response factors. Considerable variations in response factors at different scan speeds have been observed in several studies (52–54, 56). Such variations have been attributed partly to incomplete pyrolysis of substances (52). Low scan speeds may ensure complete pyrolysis; however, the life of a coated rod is generally reduced, probably due to excessive heating, when low scan speeds are used (55, 56). At low scan speeds the coefficient of variation of the reponse factors may be eleveated (56) or reduced (53) and the sensitivity of detection decreased (52, 53).

3. *Methods of Quantitation and General Applications*

In view of the variabilities in response factors of the components of a mixture analyzed by the coated-rod TLC-FID technqiue, reliable quantitative data can be obtained only if the observed peak areas are corrected by using proper calibration factors, which must be routinely determined. Use of a suitable internal standard (44, 48, 60, 61) and empirical calibration with known mixtures resembling the composition of the mixtures to be analyzed (38) are two of the techniques recommended for reliable quantitation.

Reproducibility of quantitative results obtained by coated-rod TLC-FID techniques is fairly good, at least for the major components of a mixture. For the major components, the relative standard deviations reported are in the ranges 4–6% (56), 6–13% (47, 55), and 2–10% (53). In general, the relative standard deviation seems to increase steeply with decreasing amount of the component analyzed (47, 53). Consequently, the accuracy of the quantitation of minor components of a mixture is rather low, which seems to be a limitation of this technique (53).

As pointed out earlier, the major application of the coated-rod TLC-FID technique has been in the separation and quantitation of lipid mixtures. Two typical separations of complex lipid mixtures on coated rods are illustrated in Figures 6 and 7. Industrial applications of coated-rod TLC are discussed in the next section.

IV. INDUSTRIAL APPLICATIONS OF TLC-FID SYSTEMS

The techniques of quantitative TLC with flame ionization detection have already been used in the analysis of a wide variety of industrial products. Only documented applications of these techniques are reviewed here.

A. Fats and Oils

Triacylglycerols of palm oil, shea butter, and mango kernel oil have been derivatized to their mercury(II) acetate adducts, fractionated on Chromarod-S according to the degree of unsaturation, and quantitated by FID (41). The data were in good agreement with those obtained by argentation TLC in conjunction with

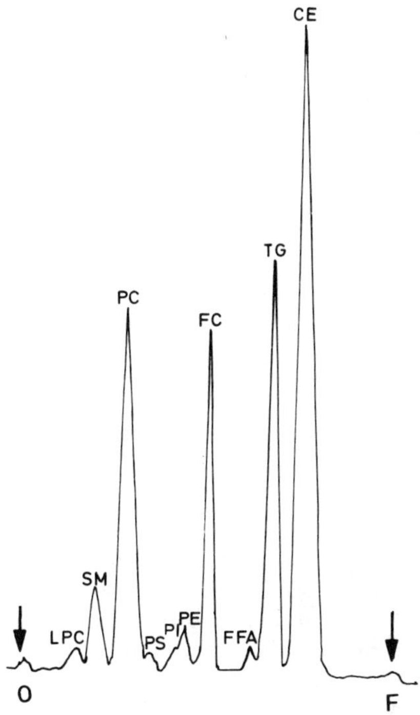

Figure 6 Chromatogram of human blood plasma lipids separated on Chromarod-SII and detected by Iatroscan TH-10. Peak identification: LPC, lysophosphatidylcholines; SM, sphingomyelins; PC, phosphatidylcholines; PS, phosphatidylserines; PI, phosphatidylinositols; PE, phosphatidylethanolamines; FC, cholesterol; FFA, unesterified fatty acids; TG, triacylglycerols; CE, cholesteryl esters. Chromatographic conditions: developing solvent: 1, chloroform/methanol/water (40:20:2.5, by vol) developed twice up to 5 cm; 2, hexane/light petroleum/diethyl ether/formic acid (30:24:6:0.08, by vol) up to the top. Detection conditions: scan speed, 3.2 cm/s; FID gases: air, 2100 ml/min; hydrogen, 180 ml/min. (Redrawn from Ref. 50 and reproduced with permission of Elsevier Publishing Co., Amsterdam, The Netherlands.)

gas chromatography. The phospholipid composition of edible oils such as corn oil, sunflower seed oil, and peanut oil, at various stages of industrial refining, has been determined by the coated-rod TLC-FID technique (51). Polar lipids, which are formed by

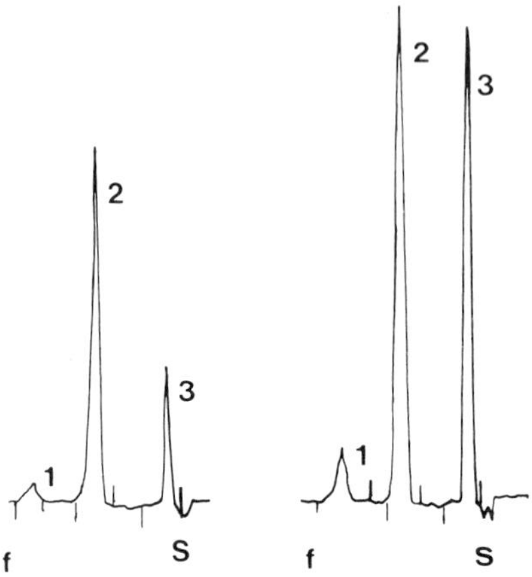

(a) (b)

Figure 7 Chromatograms of human bile separated on Chromarod-S and detected by Iatroscan TH-10. (a) sample with no visible gallstones; (b) sample with cholesterol-type gallstones. Peak identification: 1, cholesterol; 2, bile acids; 3, phospholipids; s, start; f, solvent front. Chromatographic conditions: developing solvents; 1, chloroform/light petroleum/methanol/acetone (60:20:10:10, by vol) up to about 10 cm; 2, acetone/water (50:50, by vol) up to 5 cm. Detection conditions: scan speed, 30 s per scan; FID gases; air, 2000 ml/min; hydrogen, 170 ml/min. (Reproduced from Ref. 43 with permission of Elsevier Publishing Co., Amsterdam, The Netherlands.)

oxidation of highly unsaturated triacylglycerols, have been separated from the other lipid classes of fish oils on Chromarod-S and quantitatively determined in order to assess the extent of oxidation in such oils (62). Several marine oils containing ether lipids, which are of interest to the pharmaceutical industry, have been analyzed by tubular TLC (2). Minor constituents of edible oils have also been analyzed by tubular TLC (34).

B. Fatty Acid Derivatives and Emulsifiers

Mixtures consisting of triacylglycerols and partial glycerides, such as monoacylglycerols and 1,2-(2,3-) as well 1,3-diacylglycerols, can be easily separated and quantitated by both tubular TLC (30) and coated-rod TLC (44). Such techniques have been employed to determine the composition of commercial monoacylglycerols (63, 64), which are commonly used as emulsifiers in the food industry.

Transesterification of vegetable oils with methanol yields products that are suitable as substitutes for diesel oil. The composition of such products, which contain predominantly methyl esters and minor proportions of triacylglycerols, diacylglycerols, monoacylglycerols, and unesterified fatty acids, has been determined by the coated-rod TLC-FID technique (53, 65). In a similar manner, the composition of products formed by esterification of polyhydric alcohols, other than glycerol, with fatty acids has been determined (66).

Emulsifiers obtained by condensation of long-chain fatty acids with amino alcohols, such as diethanolamine, have been analyzed by coated-rod TLC (67). Application of the TLC-FID techniques in the leather industry for the analysis of fatty acid derivatives contained in fat liquor has been described (68).

C. Food Additives

Several applications of the coated-rod TLC-FID technique have been demonstrated in the field of food additives (69). Products containing vanillin and *p*-hydroxybenzaldehyde, which are used as flavoring agents in foodstuffs, have been analyzed with good precision. Antioxidants such as butylated hydroxyanisol (BHA), dibutylated hydroxyanisole (d-BHA), and butylated hydroxytoluene (BHT) have been isolated from food products, separated on Chromarods, and quantified. Food preservatives such as benzoic acid have been extracted from fruit juices and quantitatively determined after separation on Chromarods. In some of the examples given above, the existing techniques of TLC in conjunction with spectrophotometry or gas chromatography probably have advantages compared to the TLC-FID techniques (69).

D. Pharmaceutical Products and Agrochemicals

A key paper (36) describes the application of the coated-rod TLC-FID technique in the analysis of some typical pharmaceutical, biomedical, and agrochemical products. Thus, mixtures of sulfanilic acid and sulfonamides, as well as those of common alkaloids such as quinine, codeine, brucine, strychnine, caffeine, reserpine, and aconitine, have been resolved on silica gel sintered quartz rods and detected by FID. Similarly, water-soluble vitamins of the B series have been separated from each other and from vitamin C, nicotinic acid, and nicotinamide, and subsequently detected. Mixtures of cardiac glycosides such as digitoxin, digoxin, and gitoxin, as well as genins such as digitoxingenin, digoxigenin, and gitoxigenin, have also been separated and detected without converting them to their volatile silyl ether derivatives. Steroidal hormones, i.e., estrogens, progestins, and androgens; mixtures of corticosteroids such as desoxycortisone, cortisone, and hydrocortisone; and mixtures of betamethasone, prednisone, and prednisolone have been resolved on silica gel sintered quartz rods and detected by FID. In the field of agrochemicals, an example given shows the separation and detection of a complex mixture of the following polychlorinated biphenyls and pesticides: decachlorinated biphenyl (DCB), Kanechlor (KC)-300, *p,p'*-dichlorodiphenyltrichloroethane (DDT), lindane, γ-benzenehexachloride (BHC), endrin, dieldrin, and phenylphosphonothioic acid O-ethyl O-*p*-nitrophenyl ester (NPN).

E. Coal- and Petroleum-Based Products

A major area of application of the techniques of quantitative TLC with FID has been in the analysis of high-boiling constituents of coal- and petroleum-based products. An earlier review (70) outlines the scope of the coated-rod TLC-FID technique in the semiquantitative analysis of the "hydrocarbon group types" in a wide variety of oil samples, such as atmospheric and vacuum residues, fuel oils, topped crude oils, bitumens, blown residues, pitches, and coal liquids. Such products can, in general, be fractionated on Chromarods coated with silica gel into compound types designated as saturates, aromatics, resins A, and resins B, and detected by FID. Prior to chromatography, the samples can be "deasphaltenated" by exhaustive extraction with *n*-heptane, and then both the asphaltene fraction (mainly resins) and the deasphaltenated oil (mainly saturates and aromatics) can be separately analyzed on the adsorbent-coated rods. Several variations and modifications of these techniques in the analysis of hydrocarbon group types in coal- and petroleum-based products have become known in recent years.

Characterization of bitumens, heavy oils, and synthetic fuels on the basis of their asphaltene content (the portion that is insoluble in *n*-paraffinic hydrocarbons but soluble in benzene) and their malthene content (the portion that is soluble in *n*-paraffinic hydrocarbons) is a common procedure in petroleum and allied industries (71). Asphaltenes and malthenes have been extracted from samples of bitumens, heavy oils, and distillation residues with benzene, separated on Chromarods coated with sintered silica gel, and quantitatively determined by FID (72–75). The data obtained by this technique have been found to be in good agreement (72) with those determined by a standard procedure (71), which is rather time-consuming.

Distillation residues resulting from coal liquefaction processes have been fractionated on Chromarods into component types such as malthenes (pentane-solubles), asphaltenes (pentane-insolubles, benzene-solubles), and preasphaltenes (benzene-insolubles, tetrahydrofuran-solubles) and quantitated by FID (76). Some other analyses of hydrocarbon group types, carried out by the coated-rod TLC-FID technique, include the estimation of saturates, mono- plus dinuclear aromatics, polynuclear aromatics, and polar components in the pentane-soluble malthene fraction from heavy oils and synthetic fuels (77) and in toluene extracts from coal liquefaction products (78).

Analysis of air pollution caused by coal- and petroleum-based products is another field of application of the TLC-FID techniques. Organic extracts of diesel exhaust particulates have been fractionated on Chromarod-SII into aliphatics, aromatics, and polar components and quantified by FID (79). Particulate polycyclic aromatic hydrocarbons in coal tar pitch volatiles have been separated on Chromarods from other constituents, such as lubricating oils, and determined quantitatively by FID (80).

F. Potential Future Applications

The foregoing discussion has clearly illustrated how the techniques of TLC in conjunction with flame ionization detection can be applied for the analysis of a wide variety of high molecular weight substances which are of industrial interest. For the future, one might foresee numerous other applications of the TLC-FID techniques in the fractionation and quantitation of complex mixtures, for which the other chromatographic or spectrophotometric techniques are of limited use. Choice of a method for a particular analytical problem will, of course, ultimately be governed not only by the simplicity and speed of the technique, but also by the accuracy desired for a certain set of analytical data.

Some of the potential future developments of quantitative techniques of TLC with vapor phase detectors will include the application of specific detection systems, such as the flame thermionic ionization detector (FTID) (Worldwide distributors: Newman-Howells Associates Ltd., Wolvesey Palace, Winchester, Hants, SO23 9NB England), which responds to compounds containing nitrogen or halogen atoms. Tubular TLC can be coupled with vapor phase radiation detectors (17, 22) or with a mass spectrometer as described recently (81, 82). In tubular TLC, the use of sintered adsorbent and modifications of the FID to accommodate high flow rates of the carrier gas would improve the reusability of the chromatotubes and increase scan speed. Consequently, the application of this technique could be greatly widened and the speed of analysis increased considerably.

REFERENCES

1. E. Stahl. Angew. Chem. Int. Ed. Engl. *22*, 507 (1983).
2. H. K. Mangold and K. D. Mukherjee. J. Chromatogr. Sci. *13*, 398 (1975).
3. H. K. Mangold. In: *Dietary Fats and Health,* E. G. Perkins and W. J. Visek (eds.), p. 110, American Oil Chemists' Society, Champagne, Ill., 1983.
4. U. Hezel. Angew. Chem. *85*, 334 (1973).
5. J. Sherma, J. C. Touchstone, and M. F. Dobbins. In: *Densitometry in Thin Layer Chromatography—Practice and Applications,* J. C. Touchstone, and J. Sherma (eds.), p. 95, Wiley, New York, 1979.
6. J. C. Touchstone and J. Sherma. In: *Densitometry in Thin Layer Chromatography—Practice and Applications,* J. C. Touchstone and J. Sherma (eds.), Wiley, New York, 1979.
7. O. S. Privett, K. S. Dougherty, and J. D. Castell. Am. J. Clin. Nutr. *24*, 1265 (1971).
8. T. Castellani, P. Kjelgaard-Nielsen, and J. Wolff-Jensen. J. Chromatogr. *104*, 123 (1975).
9. R. J. Komarek, R. G. Jensen, and B. W. Pickett. J. Lipid Res. *5*, 268 (1964).
10. E. Levin and C. Head. Anal. Biochem. *10*, 23 (1965).
11. E. Vioque and R. T. Holman. J. Am. Oil Chem. Soc. *39*, 63 (1963).
12. J. S. Amenta. J. Lipid Res. *5*, 270 (1964).
13. G. Rouser, A. N. Siakotos, and S. Fleischer. Lipids *1*, 85 (1966).

14. M. L. Blank, B. Verdino, and O. S. Privett. J. Am. Oil Chem. Soc. *42*, 87 (1965).

15. W. W. Christie, R. C. Noble, and J. H. Moore. Analyst *95*, 940 (1970).

16. W. W. Christie. Analyst *97*, 221 (1972).

17. K. D. Mukherjee. In: *Lipids*, vol. 2, R. Paoletti, G. Jacini, and R. Porcellati (eds.), p. 361, Raven, New York, 1976.

18. T. Cotgreave and A. Lynes. J. Chromatogr. *30*, 117 (1967).

19. A. Linenberg and O. Novick. Isr. J. Chem. *8*, 68 (1970).

20. H. P. Kaufmann and K. D. Mukherjee. Fette Seifen Anstrichm. *71*, 11 (1969).

21. E. Haahti and I. Jaakonmäki. Ann. Med. Biol. Exp. Fenn. *47*, 175 (1969).

22. E. Haahti, R. Vihko, I. Jaakonmäki, and R. S. Evans. J. Chromatogr. Sci. *8*, 370 (1970).

23. F. B. Padley. J. Chromatogr. *39*, 37 (1969).

24. J. J. Szakasits, P. V. Peurifoy, and L. A. Woods. Anal. Chem. *42*, 351 (1970).

25. J. Sjövall, E. Nyström, and E. Haahti. In: *Advances in Chromatography VI*, J. C. Giddings and R. A. Keller (eds), p. 119, Marcel Dekker, New York, 1968.

26. A. T. James, J. R. Ravenhill, and R. P. W. Scott. Chem. Ind. (London) 746 (1964).

27. J. H. van Dijk. Z. Anal. Chem. *236*, 326 (1968).

28. J. H. van Dijk. Chimia *24*, 234 (1970).

29. K. D. Mukherjee, H. Spaans, and E. Haahti. J. Chromatogr. Sci. *10*, 193 (1972).

30. K. D. Mukherjee, H. Spaans, and E. Haahti. J. Chromatogr. *61*, 317 (1971).

31. F. R. Cropper, D. M. Heinekay, and A. Westwell. Analyst *92*, 436 (1967).

32. K. D. Mukherjee. J. Chromatogr. *96*, 242 (1974).

33. O. Adam, G. Wolfram, and N. Zöllner. Fette Seifen Anstrichm. *80*, 436 (1978).

34. B. Hudalla. Doctoral thesis, University of Bonn, Bonn, Federal Republic of Germany (1978).

35. T. Laatikainen, T. Hokkanen, and E. Haahti. Scand. J. Clin. Lab. Invest. *31*, 347 (1973).

36. T. Okumura, T. Kadano, and A. Iso'o. J. Chromatogr. *108*, 329 (1975).

37. T. Okumura. J. Chromatogr. *184*, 37 (1980).

38. R. G. Ackman. Methods Enzymol. *72*, 205 (1981).

39. J. L. Sebedio, T. E. Farquharson, and R. G. Ackman. Lipids *17*, 469 (1982).

40. K. Hiramatsu and S. Arimori. J. Chromatogr. *227*, 423 (1982).

41. B. Petersson. J. Chromatogr. *242*, 313 (1982).

42. M. Foot and M. T. Clandinin. J. Chromatogr. *241*, 428 (1982).

43. M. W. Rigler, R. L. Leffert, and J. S. Patton. J. Chromatogr. *277*, 321 (1983).

44. T. Tatara, T. Fujii, T. Kawase, and M. Minagawa. Lipids *18*, 732 (1983).

45. C. C. Parrish and R. G. Ackman. J. Chromatogr. *262*, 103 (1983).

46. M. Tanaka, K. Takase, J. Ishii, T. Itoh, and H. Kaneko. J. Chromatogr. *284*, 433 (1984).

47. J. K. Kaitaranta and N. Nicolaides. J. Chromatogr. *205*, 339 (1981).

48. F. R. Farnworth, B. K. Thompson, and J. K. G. Kramer. J. Chromatogr. *240*, 463 (1982).

49. C. C. Parrish and R. G. Ackman. Lipids *18*, 563 (1983).

50. P. Mareš, M. Ranný, J. Sedláček, and J. Skořepa. J. Chromatogr. *275*, 295 (1983).

51. L. M. du Plessis and H. E. Pretorius. J. Am. Oil Chem. Soc. *60*, 1261 (1983).

52. R. T. Crane, S. C. Goheen, F. C. Larkin, and G. A. Rao. Lipids *18*, 74 (1983).

53. B. Freedman, E. H. Pryde, and W. F. Kwolek. J. Am. Oil Chem. Soc. *61*, 1215 (1984).

54. T. N. B. Kaimal and N. C. Shantha. J. Chromatogr. *288*, 177 (1984).

55. J. C. Sipos and R. G. Ackman. J. Chromatogr. Sci. *16*, 443 (1978).

56. D. M. Bradley, C. R. Rickards, and N. S. T. Thomas. Clin. Chim. Acta *92*, 293 (1979).

57. L. S. Ettre and F. J. Kabot. J. Chromatogr. *11*, 114 (1963).

58. R. G. Ackman and J. C. Sipos. J. Chromatogr. *16*, 298 (1964).

59. A. J. Sheppards, S. A. Meeks, and L. W. Elliot. J. Gas Chromatogr. *6*, 28 (1968).

60. P. van Tornout, R. Vercaemst, H. Caster, M. J. Lievens, W. de Keersgieter, F. Soetewey, and M. Rosseneu, J. Chromatogr. *164*, 222 (1979).

61. A. Martin-Ponthieu, N. Porchet, J.-C. Fruchart, G. Sezille, P. Dewailly, X. Codaccioni, and M. Delecour. Clin. Chem. *25*, 31 (1979).

62. J. K. Kaitaranta and P. J. Ke. J. Am. Oil Chem. Soc. *58*, 710 (1981).

63. E. Gantois, F. Mordret, and N. Le Barbanchon. Rev. Fr. Corps Gras *24*, 167 (1977).

64. M. Ranný, J. Sedláček, E. Mareš, Z. Svoda, and R. Seifert. Seifen Oele Fette Wachse *109*, 219 (1983).

65. B. Freedman, E. H. Pryde, and T. L. Mounts. J. Am. Oil Chem. Soc. *61*, 1638 (1984).

66. J. Poré, J.-P. Houis, and I. Rasori. Rev. Fr. Corps Gras *28*, 111 (1981).

67. J. Poré and I. Rasori. Rev. Fr. Corps Gras *31*, 23 (1984).

68. J. Poré. J. Soc. Leather Technol. Chem. *66*, 41 (1982).

69. F. Binder, P. Laugel, and M. Hasselmann. Dtsch. Lebensm. Rundsch. *75*, 111 (1979).

70. J. F. Ray, K. M. Oliver, and J. C. Wainwright. The application of the Iatroscan TLC technique to the analysis of fossil fuels. *Petroanalysis '81, IP Symposium Proceedings*, Wiley, Chichester, West Sussex, U.K., 1981.

71. *Standard Methods for Testing Petroleum and Its Products*, 36th ed., Method 143, Institute of Petroleum, London, 1977.

72. M. A. Poirier and A. E. George. J. Chromatogr. Sci. *21*, 331 (1983).

73. M. A. Poirier, N. Gueffier, J. F. Husson, and H. Bourgognon. In: *Proceedings of the International Symposium on the Characterization of Heavy Crude Oils and Petroleum Residues*, Lyon, France, 25--27 June 1984, p. 389, Editions Technip, Paris, 1984.

74. B. Sol, E. Romero, L. Carbognani, V. Sánchez, and L. Sucre. In: *Proceedings of the International Symposium on the Characterization of Heavy Crude Oils and Petroleum Residues*, Lyon, France, 25–27 June 1984, p. 394, Editions Technip, Paris, 1984.

75. J.-B. Berrut and D. Jonathan. In: *Proceedings of the International Symposium on the Characterization of Heavy Crude Oils and Petroleum Residues*, Lyon, France, 25–27 June 1984, p. 400, Editions Technip, Paris, 1984.

76. M. A. Poirier, P. Rahimi, and S. M. Ahmad. J. Chromatogr. Sci. *22*, 116 (1984).

77. M. A. Poirier and A. E. George. Energy Sources 7, 151 (1983).

78. M. L. Selucky. Anal. Chem. *55*, 141 (1983).

79. A. Obuchi, H. Aoyama, A. Otti, and H. Ohuchi. J. Chromatogr. *288*, 187 (1984).

80. H. Boden and R. Roussel. Int. Environ. Safety News 7 (June 1983).

81. L. Ramaley, M. E. Nearing, M.-A. Vaughan, R. G. Ackman, and W. D. Jamieson. Anal. Chem. *55*, 2285 (1983).

82. L. Ramaley, M.-A. Vaughan, and W. D. Jamieson. Anal. Chem. *57*, 353 (1985).

5

Analysis of Long-Chain Phenols

JOHN H. P. TYMAN *Department of Chemistry, Brunel, The University of West London, Uxbridge, Middlesex, England*

I. INTRODUCTION

A. Types of Long-Chain Phenols

The quantitative analytical applications of thin-layer chromatography discussed in this chapter are related to the long-chain alkyl phenols in natural and technical cashew nutshell liquid (CNSL) from *Anacardium occidentale*, in Japanese lac from *Rhus vernicifera*, and in poison ivy from *Rhus toxicodendron* (1). Natural CNSL, urushiol, and particularly technical CNSL are involved in a number of chemical and industrial polymerization reactions (1, 2).

Natural CNSL contains principally the phenolic acid anacardic acid (I; *n* = 0, 2, 4, 6) and cardol (II; R = H, *n* = 0, 2, 4, 6), with smaller proportions of 2-methylcardol (II; R = Me, *n* = 0, 2, 4, 6) and cardanol (III; *n* = 0, 2, 4, 6). Technical CNSL is produced industrially by the hot decarboxylation of natural CNSL (2)

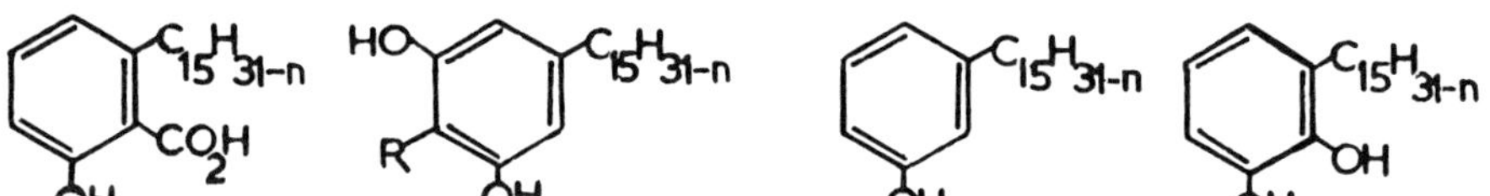

(I) (II) (III) (IV)

and consists of II (R = H, Me), III, and a small percentage of residual I. Japanese lac and poison ivy contain chiefly urushiol (IV; *n* = 0, 2, 4, 6). All the cashew phenols consist predominantly of mixtures of C_{15} unsaturated constituents with side chains of types (a), (b), (c), and (d) having *cis* double bonds at the 8 and 11 positions. In the case of urushiol as well as (a), (b), and (c) two types of triene (or 15:3) constituents are present, predominantly the 8,11,13- compound (e), accompanied by some 8,11,14- isomer (d) in Japanese lac and wholly (d) in poison ivy.

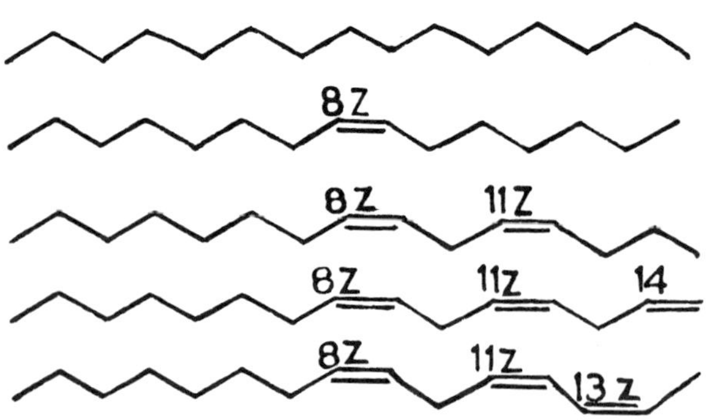

n = 0 (a)
n = 2 (b)
n = 4 (c)
n = 6 (d)
n = 6 (e)

B. Some Applications of Quantitative TLC Analysis of Phenols

1. *Nonalkyl Phenols*

Despite a considerable body of work on the TLC of phenolic materials of many types (3, 4), remarkably little quantitative work has been carried out. Phenolic acids (5), phenolic compounds (6), phenoxyalkanoic acids (7), and phenolic amines (8) have been quantitatively analyzed by a variety of different approaches. Although it is highly probable that the many qualitative separations described, e.g., those of chloro, bromo, and other phenols on silanized silica gel (9), could have been quantified, practical difficulties have inhibited this. Many phenolic compounds are important industrially, such as estrone and its relatives, the catecholamines, the phenolic antioxidants BHA and BHT, as well as the tocopherols, yet frequently applications have been qualitative.

2. *Long-Chain Alkyl Phenols*

The unsaturated constituents of anacardic acid (Ia-Id) (10), the components of natural and technical CNSL (11, 12), and those of Japanese lac (13) have all been examined by various methods of quantitative TLC. The monitoring of purification processes (14), of polymerization reactions (15), and of the solvent extraction of the natural product has enabled quantitative TLC to be used.

3. *General Observations on Quantitative TLC of Long-Chain Phenols*

Long-chain phenols have both an aromatic and an aliphatic structure typical of simpler phenolic compounds and of lipids. Thus the adsorption TLC of natural CNSL, which consists essentially of the separation of monohydric and dihydric phenols and phenolic acids according to functional group, illustrates one aspect, while the argentation TLC separation of the (15:0), (15:1), (15:2), and (15:3) constituents of each phenol has analogies in the separation of unsaturated fatty esters or glycerides. Our own work on the application of quantitative TLC was a logical development from the initial examination of the natural product by adsorption TLC followed by gravimetry, ultraviolet spectrophotometry "off the plate," and densitometry "on the plate." The qualitative approach led imperceptibly but inevitably to quantification. No such similar situation exists in liquid (or column) chromatography. Macroliquid chromatography, such as "flash chromatography" of a mixture, may have no connection with the high-performance liquid chromatography (HPLC) of similar mixtures. Many investigators, perhaps due to the enormous expansion of the technique, have proceeded directly to HPLC. Quantitative TLC was sufficiently established not to have been swept away by the avalanche of HPLC approaches, and it is a highly economical technique involving equipment which does not have to be dedicated and which has both an analytical and a preparative facility. In the ensuing sections, after an account of the general experimental methods used, some examples are given of quantitative TLC which may be applicable to other phenolic systems, since all phenols tend to be readily visualized and to be amenable to ultraviolet spectrophotometric or densitometric determination.

II. EXPERIMENTAL TECHNIQUES

A. Materials

Natural CNSL was of Mozambique or Tanzanian origin (11), as indicated in the tables, and was extracted as described (16). Technical CNSL, sometimes referred in commercial practice as raw CNSL, was obtained from Brazil (17). Urushiol was obtained from the National Industrial Research Institute, Sendai, Japan. Silica gel G (type 60) was used for the preparation for analytical or preparative plates. Later, for quantitative work, precoated commercial silica gel (F254) plates (Merck) were used. All solvents used for TLC development, analytical and preparative, were redistilled. Solvent mixtures used for development had the composition indicated in the tables.

B. Equipment

Standard TLC equipment was used throughout. For indirect spectrophotometry ("off the plate"), methanolic solutions of phenols in 1-cm cells were used with manual scanning on an SP 500 Unicam spectrophotometer between 230 and 330 nm. For constituents separated by argentation TLC, an Optica and a Perkin-Elmer model 137 spectrophotometer were used for automatic recording, but manual measurement with the SP 500 was generally found more suitable and accurate. For direct spectrophotometry and densitometry ("on the plate"), a Fisons-Vitatron flying-spot scanner (TLD 100) was used with a Fisons-Vitatron recorder/integrator. At the preliminary stage, absorption-reflectance and fluorescence modes were both examined. The apparatus contained a quartz iodine lamp (spectral range, 320–700 nm) for absorptiometry and

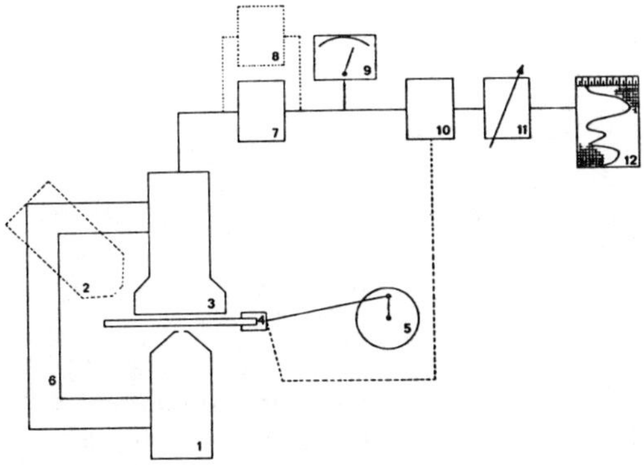

Figure 1 Arrangement of the densitometer.

a mercury lamp for fluorescence and reflectance. The various parts consist of (1) quartz iodine lamp, (2) mercury lamp, (3) photomultiplier, (4) plate holder, (5) eccentric drive, (6) optical base, (7) logarithmic amplifier, (8) linear amplifier, (9) readout meter, (10) integrator, (11) variable amplifier, and (12) recorder with integrator read-off (Figure 1). For transmission work no filter was used. The eccentricity adjustment was made so that the whole spot or band and the adjacent areas were scanned. In order to obtain peak areas with speed, use was made as far as possible of the digital readout rather than the process of "notch" counting at the baseline. The former was facilitated by having a constant baseline, since all areas above the starting baseline were automatically counted. When the direct readout was used this was started from a counting rate of one notch per second at the commencement and conclusion of the peak being examined. It was desirable to carry out integrations at a high chart speed to obtain peaks as large as possible on the chart. For the present analyses, six rolls of chart paper were used from the start to the finish of the work. Good agreement was obtained between digital readout and notch counting, which was, however, somewhat laborious. In certain cases where symmetrical peaks were obtained, triangulation could also be used. Absorption, reflectance, and fluorescence measurements were made, and the measurement of transmission with iodine-impregnated plates was found to be more satisfactory method (see Section III). Unfortunately, no opportunity was found to make a comparison with a slit-scanning type of apparatus.

C. Experimental Methods

Analytical plates (8 × 10 cm) and preparative plates (20 × 20 cm) were coated in the usual way with silica gel G with the applicator set to 0.25 and 1.00 mm, respectively. Argentation TLC was effected with silica gel G containing 15% (w/w) silver nitrate incorporated as a solution into the silica gel by slurrying. Analytical plates of all types were activated at 110°C for 1 h, while preparative plates were allowed to dry at ambient temperature overnight and then activated as before.

Samples of phenolic materials were applied by syringe in chloroform solutions. All sample applications to plates were carried out in an atmosphere of nitrogen in a nitrogen box. Chromatograms were developed in the absence of direct sunlight. For visualization, reagents and sprays were as indicated in the tables. In indirect spectrophotometry, bands visualized by 0.1% ethanolic rhodamine 6G solution were eluted with methanol. The suspensions were filtered after desorption of the phenolic material, and the filtrate was concentrated by rotary evaporation to give a residue which was freed from the indicator by washing an ethereal solution with 0.01 M aqueous sulfuric acid. After drying and evaporation, the phenolic material was examined spectrophotometrically in methanol solution. In the indirect spectrophotometry of the unsaturated constituents of anacardic acid separated by argentation TLC, the bands were visualized with 0.1% ethanolic dichlorofluorescein or 3,5-dihydroxypyrene-8,10-disulfonic acid. The former was removed by washing a petroleum ether solution of the phenolic constituent with dilute sodium hydroxide solution and the latter with aqueous washing. Silver nitrate, coeluted with the phenolic material, was removed by washing ethereal solutions with water. In direct densitometry (on the plate) after development and prior to visualization, the plates were partially freed of solvent in a nitrogen box and then placed in an oven at 100°C to remove remaining solvent and most of the ammonia (in the case of ammoniated solvents). Visualization was best achieved with iodine vapor and the plates were examined by the transmission method.

The standard solution of the component phenols used in direct densitometry contained anacardic acid (0.2371 g), cardol (0.0636 g), 2-methylcardol (0.0204 g), and cardanol (0.0224 g) made up in chloroform (10 cm^3).

In gravimetric determinations, which were used only in the exploratory stages, phenolic materials recovered by elution were freed from all extraneous material and weighed to constant weight.

III. RESULTS AND DISCUSSION

Applications of the quantitative TLC of long-chain phenols are given for the direct (on the plate) densitometric determination and the indirect (off the plate) ultraviolet (uv) spectrophotometric determination of natural and technical CNSL (Section III.A) the indirect ultraviolet spectrophotometric and gravimetric determinations of the unsaturated constituents of anacardic acid (Section III.B), and the gravimetric determination of the components of Japanese lac (Section III.C).

A. Natural and Technical CNSL

1. *Exploratory Experiments with Direct Techniques*

In preliminary experiments with self-prepared and with commercial plates, reflectance, transmission, and fluorometric procedures were examined. Fluorescence intensity measurements were complicated either by the simultaneous fluorescence and quenching of different bands or by bands of widely different colors, which required the use of filters. Measurement of transmission with sulfuric acid-charred bands was easier, although the peak areas of a given band were found to be dependent on the R_f value, the extent of charring, and the uniformity of spraying. In the case of self-prepared plates, variations in thickness of the layer and nonlinearity of area/concentration plots were major difficulties. Commercial plates were more uniform and gave linear plots, but the binder used with silica gel caused general darkening of the plate and the need for a different visualizing agent, which was found, finally, with iodine vapor.

With the cashew phenols it was essential to employ multiple development to resolve bands and to utilize the whole plate, and furthermore to develop with the least polar solvent last in the sequence to the top of the plate to avoid solvent-front interferences with integration measurements. It was found consistently that integration results depended on the development procedure, which was also found by others (18, 19), and quantitation was most accurate within a certain range of R_f values and inaccurate at high values near the solvent front. Streaking and slight tailing of bands caused inconsistency of the chart recorder baseline and was overcome by scanning the bands in a direction parallel to the baseline of the plate (horizontal scan) rather than from the bottom to the top of the plate (vertical scan). In this way all the bands of the same component were successively scanned from the unknown

Table 1 Analysis of Natural Cashew Nutshell Liquid: Summary of TLC Plates Developed, Experimental Conditions, and Observations

Plate no.	Material	Solvent	R_f[a] A	R_f[a] B	R_f[a] C	R_f[a] D	CNSL used (μl)	Standard used (μl) D	Standard used (μl) C	Observations
1	Natural CNSL (Mozambique)	(i) I[b] (ii) II	0.79	0.74	0.41	0	30	20 30 40 50	20 30 40 —	Standard commercial plate used for the series of experiments.
2		(i) II (ii) I	0.83	0.76	0.45	0	28	20 30 40 50	20 30 40 —	Standard commercial plate used for the series of experiments.
3		II only	0.95	0.95	0.66	0	30	18 30 40 —	20 30 40 —	A and B not separated.
4		(i) II (ii) I	0.83	0.66	0.37	0	30	20 30 40 50	20 30 40 50	Quite good sharp spots. Suitable for integration.
5		(i) II (ii) I	0.59	0.50	0.35	0	30	(Standard of A, B, C, D, used) 10 20 30 40 50 60		New four-component standard sepn. of A and B just sufficient for h and v scan.[c]
6		III Only	0.79	0.55	0.26	0.79	20 20 30	10 — 20 — 30 40		A and D not separated; plate suitable for h and v scan.
7		(i) II (ii) I	0.63	0.57	0.43	0	10 20 30	10 16 20 24 30 —		Sepn. of A and B not sufficient. Tailing. V but not h scan.

8		(i) I (ii) II	0.71	0.61	0.36	0	10	20	30	10	16	20	24	30	—	Sepn. of A and B. Streaking of D, spots all good size.
9		(i) II (ii) I	0.72	0.63	0.58	0	–	–	30	10	16	20	—	30	—	Reasonably good sepn. of A and B. Good plate for h and v scan.
10	Technical CNSL	(i) II (ii) I	0.82	0.52	0.24	—	—	—	30	10	16	20	—	30	—	Excellent sepn. of A and B. Bow-shaped solvent front.
11		III	0.89	0.66	0.34	—	—	—	30	10	16	20	24	30	—	Broad A spot. Too wide for scan. Bow-shaped solvent front.
12		III	0.67	0.59	0.40	—	10	20	30	—	—	20	—	40	60	Insufficient sepn. of A and B for h and v scan.
13	Natural CNSL (Tanzanian)	(i) II (ii) I	0.73	0.63	0.48	0	—	20	30	10	16	20	20	20	—	Sepn. of A and B reasonably good for h and v scan.

[a]A, cardanol; B, 2-methylcardol; C, cardol; D, anacardic acid.
[b]I, petroleum ether(b.p. 40–60°) (70)-diethyl ether (30)-NH_3; II, diethyl ether-NH_3; III, HCO_2H(2)-petroleum ether (70)-diethyl ether (30).
[c]h, horizontal scan; v, vertical scan.
Source: Reproduced from Ref. 12 by permission of Elsevier Science Publishers, B.V.

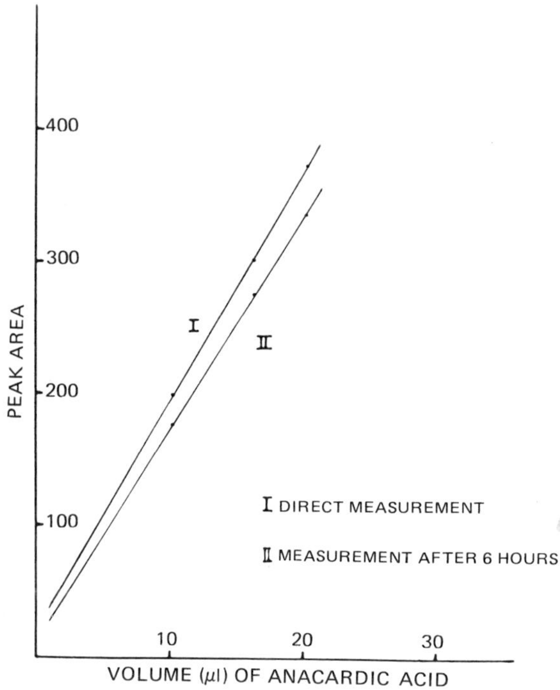

Figure 2 Effect of time on iodine visualization. I, direct measurement; II, after 6 hours.

sample and proceeding through a series of standard amounts of the same material. At the commencement of each horizontal series, the TLD instrument was rezeroed. It was also found essential to apply the calibration standard from the same solvent as the unknown sample in approximately the same sample magnitude. With these precautions, a plot of peak area versus amount of component phenol was found to be linear. Nonlinearity was consequent upon patchy or inhomogeneous charring due to persistent residual effects of ammonia in the developing solvent, and the inherent difficulty of uniform manual spraying was overcome by vapor impregnation with bromine and, more usefully, iodine.

2. *Visualization with Iodine and Quantitation by Densitometry*

Iodine has been little investigated as a visualizing agent in quantitative work, and although it has been commented (20) that it is

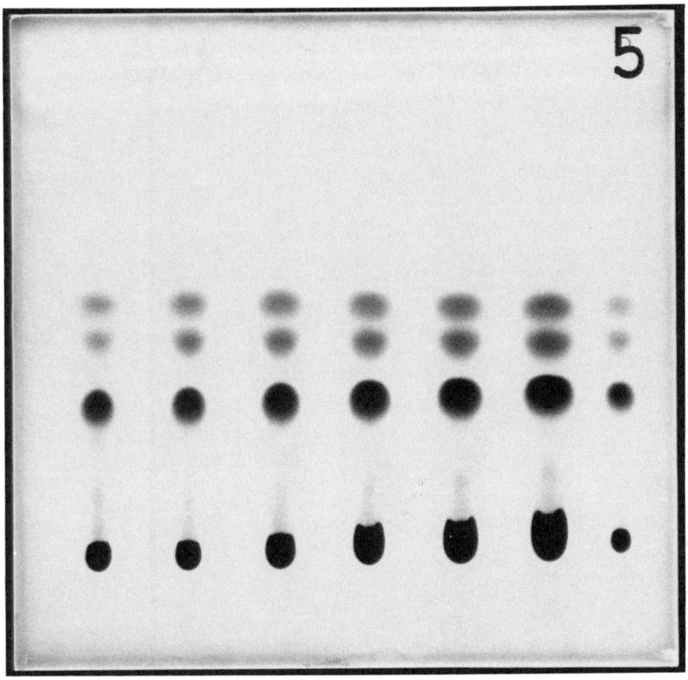

Figure 3 Micrograph of TLC plates, 5–13.

suitable for compounds possessing no absorption in the UV or visible region, we have found it to be highly satisfactory for compounds having UV absorption but little color. A broad-wavelength absorption occurs much as with sulfuric acid charring, and it seems likely (21) that iodine partialy interacts with the ring. The reaction of iodine with other materials has been investigated (22). Absorption measured on freshly impregnated plates and those from which iodine had partly volatilized still showed linearity (Figure 2). Nevertheless, it was found convenient to superimpose a blank glass plate over that being examined and seal the edges with "selotape."

The results of a number of chromatographic experiments with natural and technical CNSL are summarized in Table 1. Following an exploratory series of five plates, the first four listed in the table were developed with the two principal components anacardic acid and cardol and visualization effected with iodine. For plates 5–13 a combined calibration standard of anacardic acid, cardol,

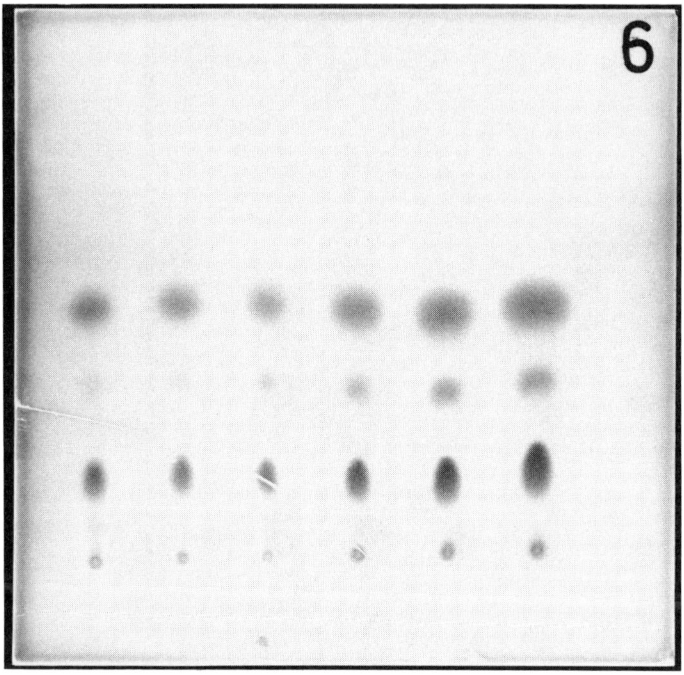

Figure 3 (Continued)

2-methylcardol, and cardanol was used. With solvents I and II, multiple development was used to resolve the components, although because cardanol and 2-methylcardol were present as comparatively faint spots until visualization, there was some trial and error in the use of solvent I, the less polar second solvent. The results obtained on iodine visualization are shown in nine photographs (Figure 3). Of these nine plates, six were suitable for integration, namely plates 5, 6, 8, 9, 10, and 13. In plates 7 and 12 the upper two bands were too close and in plate 11 the upper band was too large for the scanner (beyond a certain band size the eccentricity adjustment of the instrument could not cover both the background and the band) and was an unresolved mixture of anacardic acid (D) and cardanol (A). In plates 6 and 10 the upper band was an unresolved mixture of anacardic acid and cardanol. Plate 10 was less satisfactory because of the bow-shaped solvent front, which led to a requirement for a greater scanning area than was usually

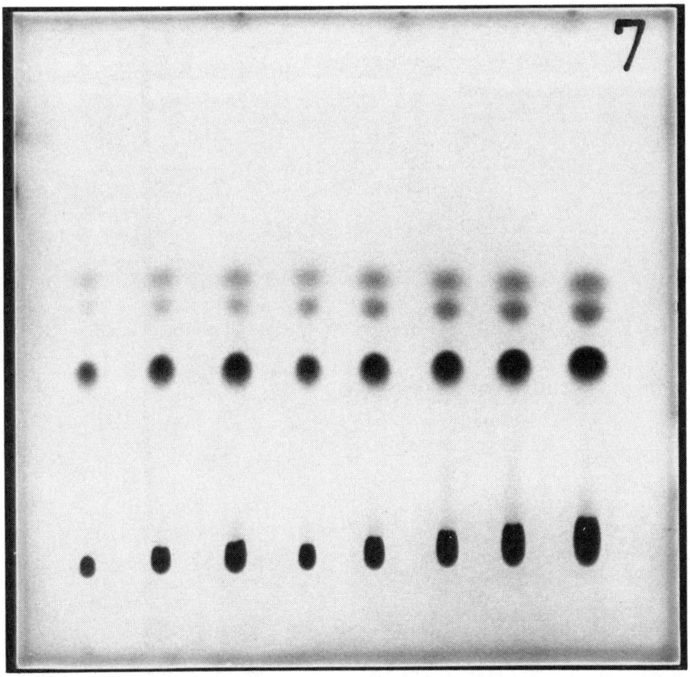

Figure 3 (Continued)

necessary. The integration results for three of the better plates, 8, 9, and 13, are given in Table 2.

In general, it was more satisfactory to use solvents I and II, in which anacardic acid remained at the baseline as the ammonium salt, rather than solvent III, in which the free acid behaved as a nonpolar intramolecularly hydrogen-bonded material (11). Either resolution from cardanol was insufficient or the bands were too large for the scanner with both technical CNSL (low D/A ratio) and natural CNSL (high D/A ratio). The objective in using solvent III had been to separate all the component phenols from polar polymeric material, which normally hardly migrated from the baseline, and this aspect requires further investigation. The two sets of results for plates 8 and 9 were obtained after an elapse of time in each case, and the areas were determined to find the effect of a slight loss of iodine at the visualization stage. Figure 4 shows plots of peak area vs. volume of standard solution without a least-squares

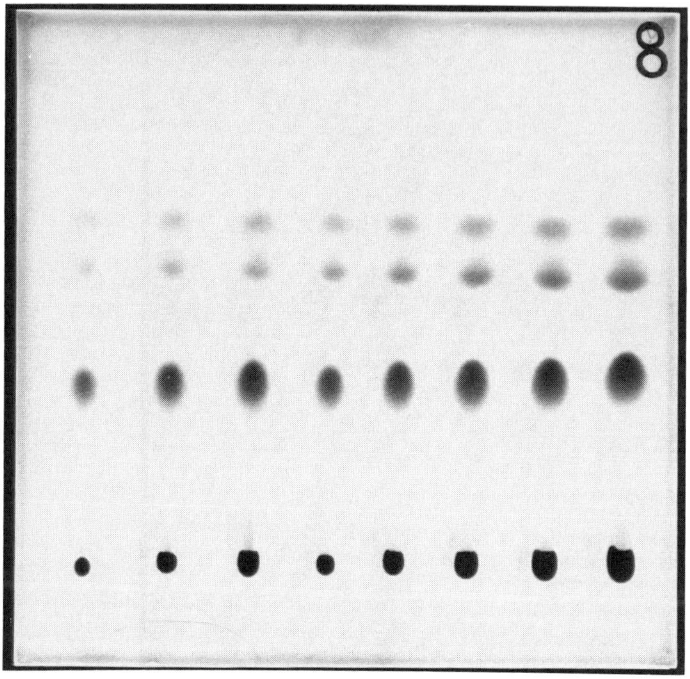

Figure 3 (Continued)

treatment. Since the linear calibration plots usually passed through the origin, application of the least-squares method to determine the slope in the form $m = \Sigma(xy)/\Sigma(x^2)$ was considered justified. From plots of the slope m for the four component phenols, anacardic acid, cardol, 2-methylcardol, and cardanol, with the axes representing integrated area and volume (microliters) of the four standards, the volume of unknown CNSL corresponding to the area measured for the particular component was found. From these equivalent volumes the composition of the natural CNSL in terms of the relative weights of the four principal components was readily calculated. If the volume with respect to anacardic acid (0.2371 g present in the standard) was found to be x μl and the volume with respect to cardol (0.0636 g in the same standard) was y μl, then the relative amount of cardol present was (0.0636 g × y/x), and similarly for the other components: 2-methylcardol (z μl present; 0.0204 g × z/x) and cardanol (w μl present; 0.0224 g ×

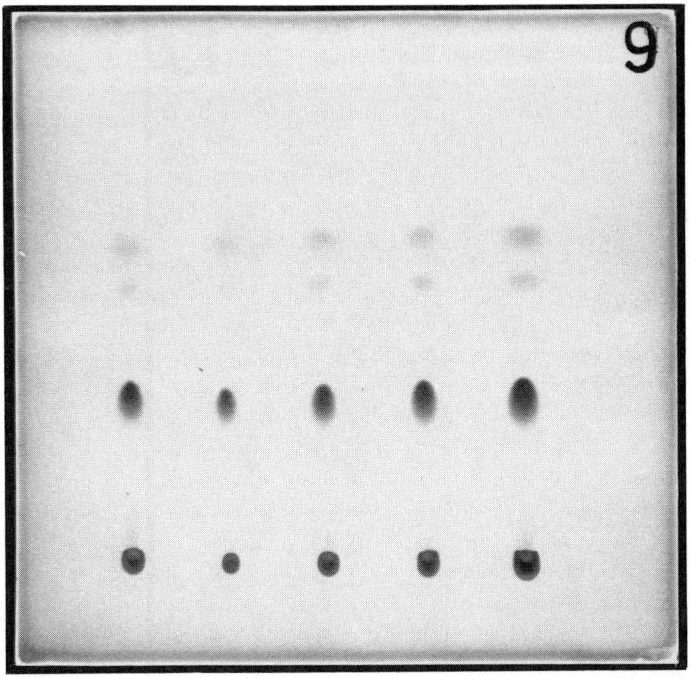

Figure 3 (Continued)

w)/x). These relative weights were used to obtain normalized results. Table 3 shows the percentage composition of component phenols in the natural CNSL sample examined. From the total weight found the percentage recovery was calculated and is given in the final column. The results for plate 10 (technical CNSL) could be similarly processed, but have not been included in the present series because of the desirability of separating the top mixed band.

In a detailed and more theoretical study with a similar densitometer (23) it was concluded that corrections for background fluctuation and light scattering should be incorporated. The present results suggest that, with uniform commercial plates, these two factors may be of less importance in the practical procedure adopted in this work, but become more significant for minor compounds and those of high R_f value. The present procedure of assuming "y = mx" plots approximates a correction for irrelevant absorption.

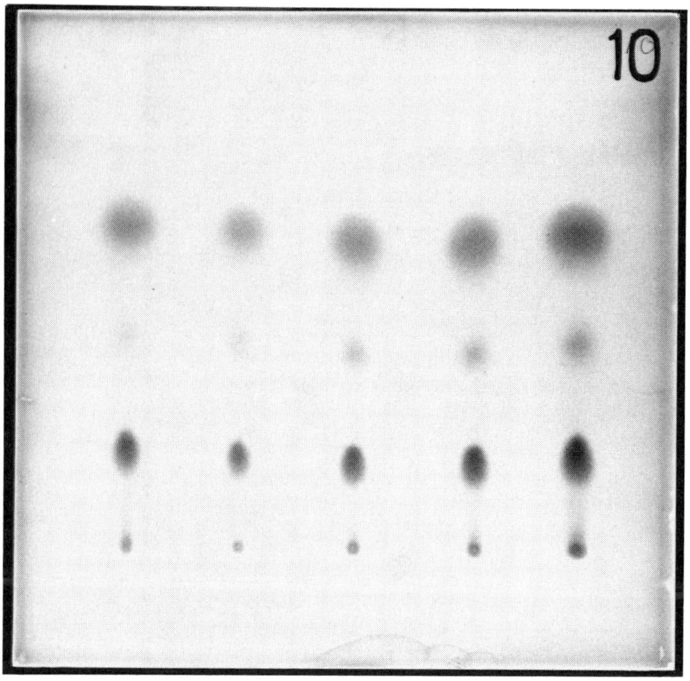

Figure 3 (Continued)

3. *Quantitation by Indirect Ultraviolet Spectrophotometry*

Experiments on the elution of the component phenols from the TLC plate followed by UV spectrophotometry (11) stemmed from attempts to use gravimetry for quantitation. It was desirable to have some characterization of the eluted bands since the total recovered material from the elution was greater than 100%. After spectrophotometric determination, two forms of correction were used. Initially, the epsilon values and λ(max) of the purified component phenols were determined. The UV absorption maximum measured is attributable to the substituted phenolic ring, and the unsaturated side chain does not absorb and interfere in the region of this absorption maximum. The purity of materials recovered from the natural CNSL being analyzed was expressed as a percentage of the component phenol standard. Second, to allow for background absorption (24) which could not be ignored, a corrected optical density

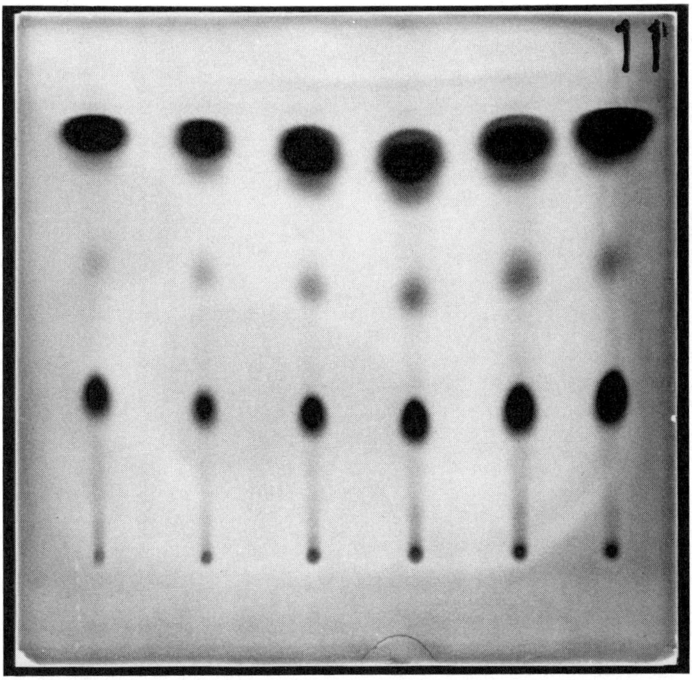

Figure 3 (Continued)

was used for determination of the component phenols. The absorption spectra of the four component phenols are shown in Figure 5. The nature of the correction is shown in Figure 6. The optical density X^1 at $\lambda(\max)_X$ was adjusted to $X^1 - (A^1 + B^1)/2$, where A^1 and B^1 are the optical densities at λ_A and λ_B. Although insufficient results have been obtained to determine the accuracy of the indirect spectrophotometric procedure, the average results of two determinations which themselves showed good agreement are given in Table 4. They differ from the results obtained by the direct method, although it must be remembered that the latter is based on peak area while the indirect method is based on peak height (optical density or absorbance) at the absorption maximum, the traditional procedure in UV spectrophotometry. Area determination is less straightforward, but could clearly be of more relevance where the absorption curves are not symmetrical. Derivation of response factors in terms of area per gram by use of

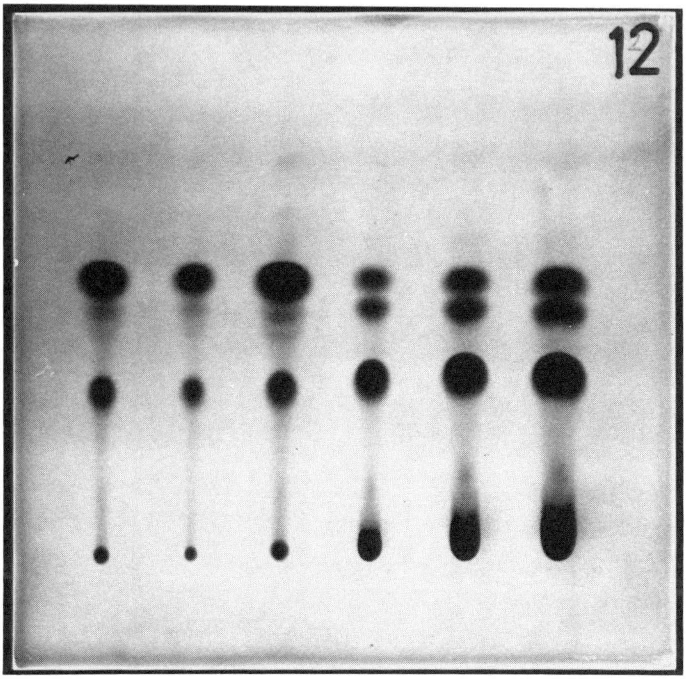

Figure 3 (Continued)

standard solutions of pure component phenols and their application to the peak areas of the material separated by elution would probably be a more legitimate way of comparing the direct densitometric and the indirect spectrophotometric procedures.

B. Unsaturated Anacardic Acid

1. *Quantitation by Indirect Spectrophotometry*

It was expected that by the use of an initial TLC separation of the component phenols of natural and technical CNSL, argentation TLC of the individual phenols by silica gel G would lead to a quantitative analysis of the (15:0), (15:1), (15:2), and (15:3) constituents. This procedute worked well for the main component of natural CNSL, anacardic acid, although it was less satisfactory for cardol, 2-methylcardol, and cardanol because of some minor

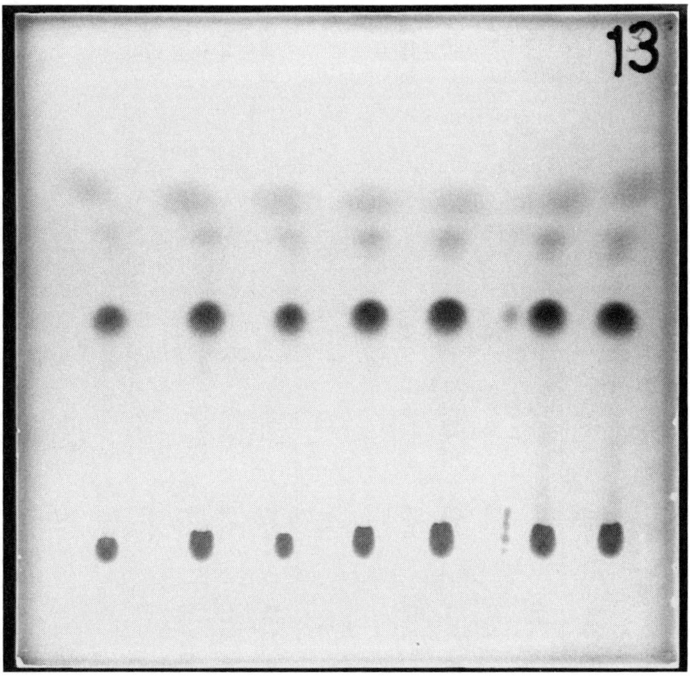

Figure 3 (Continued)

oxidative side reactions. For qualitative purposes with the latter phenols the argentation TLC method was invaluable (11).

The constituents of anacardic acid were best separated in petroleum ether/diethyl ether/formic acid mixtures, and from a number of exploratory runs, solvent system 5 shown in Table 5 was found to be most convenient (10). Silica gel G containing 10% silver nitrate was less effective than gels containing 15 or 20% silver nitrate, and the latter was used in the separation of the (15:0), (15:1), (15:2), and (15:3) constituents on preparative-type plates. It was found useful to elute the silica gel with methanol and to remove the coextracted silver nitrate from the methanolic extract concentrated to a smaller volume but not to dryness. Each constituent gave a single band with a different R_f value on $AgNO_3$ analytical plates, while on unimpregnated plates they all had the same R_f. Each separated constituent made up to 25 cm^3 (in methanol) was examined by UV spectrophotometry from 240 to 330 nm. Materials from the solvent front and from the baseline showed no

Table 2 Analysis of Natural Cashew Nutshell Liquid by TLC/Densitometry*

Plate no.	Component phenol	Volume (μl) of: Unknown CNSL and peak areas			Standards and peak areas					Slope (M)
		10	(i) 20	(ii)† 30 μl	10	16	20	24	30 μl	
8	Anacardic acid	125	205	296	180	270	330	401	513	1.3546
	Cardol	187	291	388	253	371	430	497	—	1.7414
	2-Methylcardol	—	128	185	201	323	422	535	—	1.7092
	Cardanol	256	272	345	309	505	610	713	—	2.4314
		(i)a 10	(ii)b 20	(iii)c 30 μl	10	16	20	24	30 μl	
8	Anacardic acid	97	176	255	153	227	294	343	440	1.1639
	Cardol	135	249	334	219	315	367	434	512	1.4461
	2-Methylcardol	61	98	147	146	247	331	414	522	1.3487
	Cardanol	122	172	244	205	370	429	514	615	1.6966
		—	—	(i)a 30 μl	10	16	20	—	30 μl	
9	Anacardic acid	—	—	270	178	277	337	—	451	1.2793

	Cardol	—	—	273	160	255	324	—	446	1.2338
	2-Methylcardol	—	—	141	64	172	375	—	419	1.1334
	Cardanol	—	—	243	158	284	384	—	472	1.3509
		—	—	(i)b 30 μl	10	16	20	—	30 μl	
9	Anacardic acid	—	—	336	199	303	375	—	510	1.4318
	Cardol	—	—	315	186	300	381	—	519	1.4420
	2-Methylcardol	—	—	171	136	273	372	—	648	1.1573
	Cardanol	—	—	211	114	297	410	—	609	1.6208
		—	(i) 20	(ii) 30 μl	10	16	20	20	20 μl	
13	Anacardic acid	—	227	322	192	297	350	354	410	1.4912
	Cardol	—	270	383	290	424	479	443	506	2.0322
	2-Methylcardol	—	218	351	282	466	588	662	672	2.4432
	Cardanol	—	665	889	580	889	1022	—	—	4.2819

*Volumes of CNSL and standard solutions, peak areas and slopes of calibration curves.
†The headings (i) etc. refer to the third column in Table 3.
Source: Reproduced from Ref. 12 by permission of Elsevier Science Publishers, B.V.

Table 3 Percentage of Composition of Component Phenols in Natural Cashew Nutshell Liquid by (on the Plate) TLC/Densitometry

Vol. of CNSL (μl)	Plate no.	Expt. no.	Equiv. μl				Wt. component phenol found			
			Anacardic acid	Cardol	2-Methyl-cardol	Cardanol	Anacardic acid	Cardol	2-Methyl-cardol	Cardanol
20	8	(i)	12.16	13.20	5.88	8.76	0.1442	0.0420	0.0060	0.0098
30		(ii)	17.44	17.92	8.60	11.36	0.1378	0.0380	0.0058	0.0085
10		(i)a	6.6	7.36	3.76	6.56	0.1565	0.0468	0.0077	0.0147
20	8	(ii)b	12.08	13.6	6.24	9.04	0.1432	0.0432	0.0064	0.0101
30		(iii)c	17.44	18.24	9.36	13.00	0.1378	0.0387	0.0064	0.0097
30	9	(i)a	16.96	17.76	10.08	14.24	0.1340	0.0376	0.0106	0.1892
30	9	(i)b	18.64	18.24	9.8	11.8	0.1473	0.0387	0.0067	0.0088
20	13	(i)	12.24	10.63	7.2	12.48	0.1451	0.0038	0.0073	0.0140
30		(ii)	17.28	15.14	11.52	16.64	0.1366	0.0321	0.0078	0.0124

Source: Reproduced from Ref. 12 by permission of Elsevier Science Publishers, B.V.

Vol. of CNSL (μl)	Total wt.	Composition (%)				$\% \left(\frac{\text{CNSL found}}{\text{weight taken}}\right)$
		Anacardic acid	Cardol	2-Methyl-cardol	Cardanol	
20	0.2020	71.38	20.79	2.97	4.86	102.6
30	0.1901	72.48	19.98	3.07	4.46	96.6
10	0.2257	69.35	20.74	3.39	6.51	—
20	0.2029	70.57	21.31	3.13	4.99	103.1
30	0.1926	71.57	20.08	3.30	5.04	97.8
30	0.1892	70.85	19.90	3.62	5.62	96.1
30	0.2015	73.13	19.19	3.30	4.37	102.3
		Av. 71.33	20.28	3.25	5.12	99.75
		±1.25	±0.71	±0.22	±0.74	
20	0.2002	74.46	16.89	3.67	6.98	93.7
30	0.1889	73.30	16.99	4.15	6.58	—
		Av. 72.38	16.94	3.91	6.78	Av. 98.9

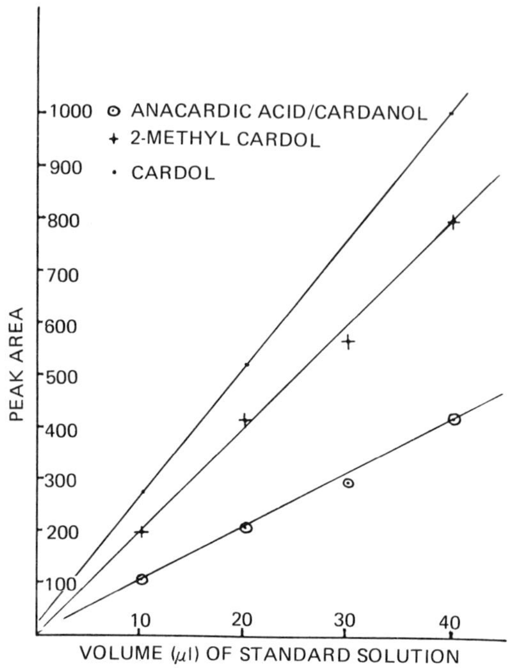

Figure 4 Plate 6. Peak area (from densitometer) versus volume of standard solution.

absorption maxima, while each constituent had a UV maximum at 308–309 nm. The optical densities averaged from three TLC determinations are shown in Table 6. Since only the molar extinction coefficients are uniform for the three unsaturated constitutents, the optical densities recorded were corrected by using molecular weights 346 and 344 for the monoene and diene, respectively, and the relationship concentration (c) = (optical density × molecular weight)/(molar extinction coefficient × 10).

The correction is very minor. The averages of the results obtained were: (15:3) constituent, 45.1%; (15:2) constituent, 16.3%; and (15:1) constituent, 38.7%. A very small proportion of the saturated constituent was believed to be present in the monoene fraction, since elution of a series of strips from above the monoene band to the top of the plate failed to yield any material.

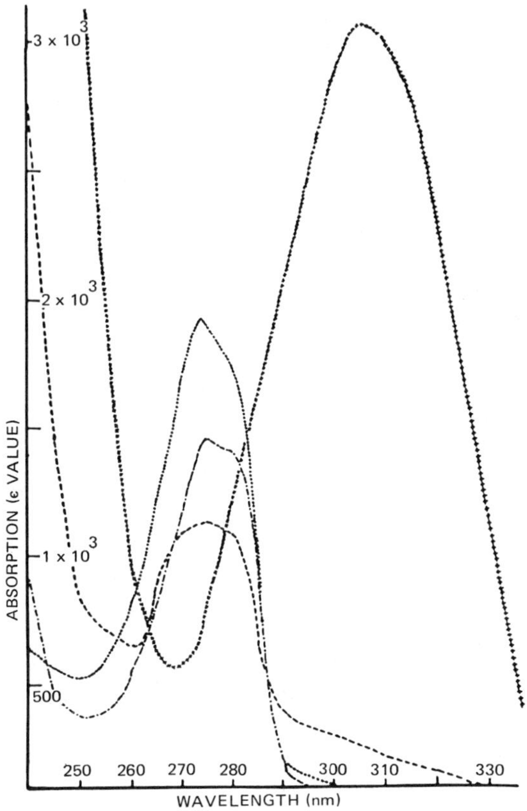

Figure 5 Ultraviolet absorption spectra of the four component phenols (methanol solution) from natural CNSL of Tanzanian origin. ♦♦♦, Anacardic acid; . . . , cardanol; •—•, cardol; — — —, 2-methylcardol.

2. *Quantitation by Gravimetry*

During a series of separations of anacardic acid a total of 11 preparative plates impregnated with 20% silver nitrate were developed in solvent 5. The combined bands after visualization and extraction as described in the experimental and preceding section gave the (15:3) constituent, 0.0889 g (44.1%); the (15:2) constituent, 0.0352 g (17.4%); and the (15:1) constituent, 0.0774 g (38.4%), thus indicating a good measure of agreement with the spectrophotometric procedure.

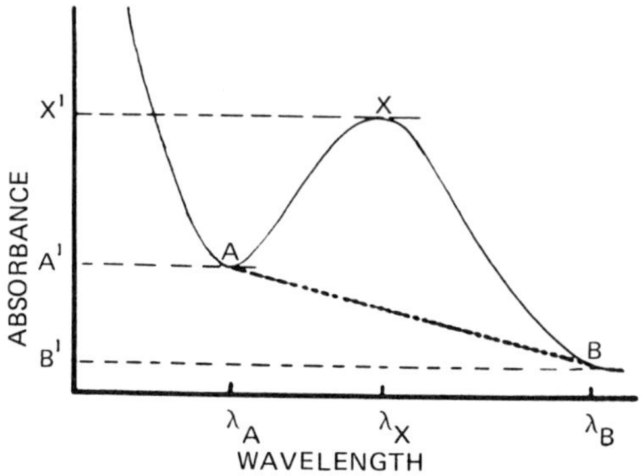

Figure 6 Typical correction procedure, used in UV indirect spectrophotometric method for determining irrelevant background absorption. For details, see text.

C. Japanese Lac

Although our own work on the quantitative analysis of Japanese lac is incomplete, the increasing use of this ancient article of commerce as a component in surface coatings (1) makes its quantitative compositional study of interest. Following initial experiments with a variety of solvents (13), chloroform-ethyl acetate (90:10) was found valuable for the separation of urushiol (IV; a-e), a small proportion of cardanol, hydroxyurushiol, dehydro "dimeric" components, and polymeric material.

From four preparative TLC separations followed by elution and workup, the composition was found by gravimetry to be urushiol (61.9%), cardanol (3.3%), hydroxyurushiol (3.0%), dehydro "dimeric" substances (4.7%), and polymeric material (27.1%). The streaking and manipulative problems with this labile material led us to separate the hydrogenated product. Although less streaking was encountered, delineation of bands at the visualization stage was more difficult and considerable variations were found from plate to plate. The unsaturated and hydrogenated products were in general more conveniently examined by gas-liquid chromatography (GLC) of the derived mono- and dimethyl ethers (13). These methods confirmed the TLC findings and are discussed in the next section.

Table 4 Determination of Percentages of Component Phenols in Natural Cashew Nutshell Liquid[a] by (Elution) TLC and UV Spectrophotometry (off the Plate)

Component phenol	λ(max) (nm)	ε[b] found	ε pure (av.)	Purity (%)	Fraction[c] wt. (g)	Fraction wt. × purity (1)	Optical[d] density (O.D.) correction factor (2)	(1) × (2)	Composition (%) (normalized)
Anacardic acid	306	2914	2998	97.19	0.0924	0.0898	0.919	0.1626	74.64
		2907		96.96	0.0898	0.0871			
Cardol	277	1400	1545	90.65	0.0512	0.0464	0.904	0.0419	19.25
2-Methylcardol	275	882.3	1100	80.18	0.0083	0.0066	0.546	0.0036	1.65
Cardanol	275	1653	1893	87.29	0.0133	0.0116	0.837	0.0097	4.46

[a]Material of Tanzanian origin was used.
[b]E (1%, 1 cm) × molecular wt. (MW) × 10^{-1}. Anacardic acid (MW 344), cardol (MW 314), 2-methylcardol (MW 328), and cardanol (MW 300).
[c]Only the four components have been included and minor materials ignored.
[d]

$$\left[\mathrm{O.D.}_{\lambda\max} - \left(\frac{\mathrm{O.D.}_{\lambda a} + \mathrm{O.D.}_{\lambda b}}{2}\right)\right] \Big/ \mathrm{O.D.}_{\lambda\max}.$$

Source: Reproduced from Ref. 12 by permission of Elsevier Science Publishers, B.V.

Table 5 R_f Values of the Constituents of Anacardic Acid

Solvent system	Solvent proportions			R_f values		
	Petroleum ether	Diethyl ether	Formic acid	Monoene	Diene	Triene
1	50		1		All at baseline	
2	50		3		All at baseline	
3	50	10	1		Poor separation	
4	50	10	3		Poor separation	
5	50	20	3	0.69	0.51	0.33
6	50	20	5	0.69	0.52	0.33
7	50	30	3	0.90	0.75	

Table 6 UV Absorption of the Constituents of Anacardic Acid from Natural Cashew Nutshell Liquid

Optical density observed			Optical density corrected		Total optical density	Total optical density (%)		
Triene (T)	Diene (D)	Monoene (M)	Diene	Monoene	(T + D + M)	Triene	Diene	Monoene
0.581	0.180	0.498	0.181	0.504	1.266	45.9	14.3	39.8
0.995	0.408	0.842	0.410	0.852	2.257	44.1	18.2	37.8
0.987	0.356	0.830	0.357	0.840	2.184	45.2	16.4	38.5

Source: Reproduced from Ref. 10 by permission of Elsevier Science Publishers, B.V.

IV. COMPARISON OF QUANTITATIVE TLC WITH OTHER CHROMATOGRAPHIC METHODS

A. Natural CNSL

Quantitative TLC, when operated in the adsorption mode, is effective in separating solutes in mixtures according to their functional groups and is almost without effect on the individual members of a homologous series. By comparison, of course, GLC and HPLC (depending on the solvent choice) invariably give total separations. While from the aspect of complete analysis this is highly desirable, there is little doubt that quantitative TLC, particularly in cases where knowledge of homology is less important, gives very direct and compact compositional information.

In our laboratory we have experienced remarkable agreement between GLC, densitometric methods, and indirect UV spectrophotometric methods. Even gravimetry shows a reasonable measure of agreement with these three approaches. Table 7 gives compositional results for natural CNSL of Mozambique and Tanzanian origin. Agreement is generally good for the two major components, but with the minor materials 2-methylcardol and cardanol there is some divergence, possibly attributable to relative background absorption and light scattering effects.

Both natural CNSL and technical CNSL contain polymeric material, which has been determined in the former by HPLC (25) and in the latter by GLC (26) and HPLC (17). The objective of solvent studies, particularly with solvent III, in the development and resolution of the component phenols of natural and technical CNSL had been to locate and determine polar polymeric material. The material left at the baseline in plates 10, 11, and 12 proved difficult to quantify by densitometry and was more amenable to quantification by elution and UV spectrophotometry. By contrast, HPLC analysis was more straightforward. In all the chromatographic methods, however, the problem of relating the polymer found to a characterizable polymer standard is inherently more difficult that with the analyses of the well-defined principal component phenols, each of which can be obtained pure in a standard calibration solution.

The HPLC method for natural CNSL (25) can be compared with the densitometric method if the percentage of the four principal component phenols is taken and the percentage of polymer is ignored. These results are shown in Table 8, which gives the results for three different chromatographic methods of analysis, namely HPLC, densitometry, and GLC. The percentages of major components agree excellently, and a discrepancy arises only with 2-methylcardol.

Table 7 Comparison of Different Analytical Methods for Percentage Composition of Natural CNSL

Source	Method[a]	Composition (%)			
		Anacardic acid	Cardol	2-Methylcardol	Cardanol
Mozambique	UV densitometry	71.37	20.28	3.25	5.12
		±1.25	±0.71	±0.22	±0.74
	GLC	71.51	22.34	2.78	3.37
		±0.23	±0.11	±0.72	±0.11
Tanzania	UV densitometry	72.38	16.94	3.90	6.70
	GLC	74.01	15.02	1.74	9.23[b]
		±0.68	±0.51	±0.24	±0.09
	UV indirect	74.61	19.25	1.65	4.46
	Gravimetric	71.99	19.94	2.98	5.09

[a]UV densitometric, UV indirect, and gravimetric methods determine, in addition to C_{15} components, the C_{17} and traces of C_9, C_{11}, C_{13}, while GLC simply determines C_{15} components.
[b]This sample of material has been kept over a period of 6 years between GLC and the other analytical methods. Some decarboxylation of anacardic acid, resulting in a higher percentage of cardanol, has occurred.
Source: Reproduced from Ref. 12 by permission of Elsevier Science Publishers, B.V.

Table 8 Comparison of Percentages of Component Phenols in Mozambique Natural CNSL Determined by Different Chromatographic Methods

Method	Anacardic acid*	Cardol	Cardanol	2-Methylcardol
HPLC	(a) 69.29	22.87	6.77	1.07
	(b) 71.19	22.87	5.12	1.07
	(c) 71.65 ± 0.493	22.17 ± 0.398	5.13 ± 0.374	1.08 0.049
Densitometry	71.37 ± 1.25	20.28 ± 0.71	5.12 ± 0.74	3.25 ± 0.74
GLC	71.51 0.23	22.34 ± 0.11	3.27 ± 0.11	2.75 ± 0.72

*(a) Normalized percentage composition for the four components calculated from Table 4 (Ref. 25); (b) adjustment of the percentage anacardic acid for that decarboxylated to cardanol; for the Mozambique sample the percentage cardanol (5.12) has been used as the basis for the calculation to compare with the densitometric value. The difference (6.77−5.12) × 344/300 is added to the percentage anacardic acid, 69.29; (c) normalization of (b).

Table 9 Comparison of Compositional Results for the Unsaturated Constituents of Anacardic Acid Determined by Different Techniques

Method	Source material	Composition (%)		
		Monoene (15:1)	Diene (15:2)	Triene (15:3)
TLC/UV spectrophotometry	Anacardic acid	38.7	16.3	45.1
Gravimetry	Anacardic acid	38.4	17.4	44.1
GLC[a]	Methyl anacardate	39.9	18.0	42.1
MS[a]	Methyl anacardate	38.4	17.4	44.1

[a]These results have been obtained by normalizing the percentage results for the (15:1), (15:2), and (15:3) constituents.

B. Anacardic Acid

The analyses of the constituents of anacardic acid by TLC/UV spectrophotometry and by gravimetry have been compared with the later results obtained by GLC (27) and by mass spectrometry (MS) (28) with methyl anacardate. The results are shown in Table 9. Despite the use of material obtained from different sources and in the form of the methyl ester in the more recent work, the agreement is reasonably good between the four procedures. It seems likely that the percentage of (15:0) constituent in the anacardic acid used in the TLC experiment was quite low. Although the GLC and MS methods require an initial TLC separation of the component phenols and a subsequent derivatization step, the TLC/UV approach is free from this last requirement. All the methods require the availability of pure component phenols and their pure unsaturated constituents.

C. Japanese Lac

Relatively little quantitative work by alternative methods has been carried out with this material. Generally, dark-colored phenolic materials such as Japanese lac and technical CNSL present difficult problems with optical methods, largely due to streaking and tailing effects at the TLC stage. A GLC examination of trimethylsilylated hydrogenated Japanese lac (29) with (15:0)-cardanol as an internal standard and (15:0)-urushiol as the component phenolic reference material, both as trimethylsilyl derivatives, indicated that 74.5% of urushiol and probably nearly 25.5% of polymerizable substances were present. These results show a measure of agreement with the gravimetric method, but clearly much remains to be done in this area. Progress has been made with HPLC with the natural product either derivatized as the diacetate (30) or in the free phenolic form (31).

V. CONCLUSIONS

The examples given in this chapter of the application of quantitative TLC to long-chain phenol chemistry have indicated that the technique can very nearly give a total analysis of the component phenols of phenolic lipids and, in the case of natural CNSL, compositional information on the unsaturated constituents of certain of the component phenols. By further improvements it is believed that quantitative analysis of the unsaturated constituents of cardol, 2-methylcardol, and cardanol could be effected. To make the method comprehensive in the analysis of homologous component phenols, partition rather than adsorption TLC would be required. Some progress has been made in this direction (32). The determination of polymeric material in phenolic lipids remains a problem which requires solution before total analysis is achieved. The investigation of other adsorbents and solvent systems is likely to bring improvements. The TLC plate for densitometry is a permanent record, and rather than being regarded as static it may lend itself to computer graphics processing and perhaps lessen the present requirement for extensive TLC expertise. Although total analyses have been achieved both by GLC (26) and by HPLC (17, 25), these methods often require a comprehensive outlay on equipment and dedication to that objective. By contrast, quantitative TLC is inexpensive. It is a valid and useful method which requires patience and considerable expertise with TLC analysis, and for this reason it may sometimes have been prematurely abandoned in favor of more instrumentated techniques. Both indirect and direct procedures have an application. The former, involving elution of bands and spectrophotometric determination, marked the start of quantitative TLC (33–41). Direct determination by densitometry or fluorimetry (42-54) has led to theoretical, instrumental, and computing aspects. Some of these may offset the traditional view of TLC as an inexpensive and rapid method of analysis. In the present account emphasis has been placed on practical aspects of achieving success with quantitative TLC and on a comparison of results with those from other chromatographic techniques. The use of alternative forms of support and the combination of TLC with other techniques such as GLC (54) offer other approaches to quantification. Nevertheless, in the reviewer's opinion, the simplicity of quantitative TLC is sufficient to ensure that it will always remain an indispensable procedure at some stage in many investigations of natural and synthetic products.

REFERENCES

1. J. H. P. Tyman. Chem. Soc. Rev. *8*, 499 (1979).
2. J. H. P. Tyman. Chem. Ind. (London) 59 (1980).
3. E. Stahl (ed.). *Thin Layer Chromatography, A Laboratory Handbook*, , Allen & Unwin, London, 1969.
4. J. C. Touchstone and M. E. Dobbins. *The Practice of Thin Layer Chromatography*, 2d ed., Wiley, New York, 1983.
5. H. Schmidtlein and K. Herrmann. J. Chromatogr. *115*, 123 (1975).
6. E. Ragazzi and G. Veronese. J. Chromatogr. *77*, 369 (1973).
7. K. S. Brenner, K. Müller, and P. Satel. J. Chromatogr. *90*, 382 (1974).
8. N. Seiler. J. Chromatogr. *63*, 97 (1981).
9. L. Lepri, P. G. Desideri, and D. Heimler. J. Chromatogr. *248*, 308 (1982).
10. J. H. P. Tyman and N. Jacobs. J. Chromatogr. *54*, 83 (1971).
11. J. H. P. Tyman and L. J. Morris. J. Chromatogr. *27*, 287 (1967).
12. J. H. P. Tyman. J. Chromatogr. *166*, 159 (1978).
13. J. H. P. Tyman and A. J. Matthews. J. Chromatogr. *235*, 149 (1982).
14. J. H. P. Tyman, M. Patel, and A. Manzara. UK Patent App. GB 2066820A (1981).
15. A. J. Hawkes, A. A. Durrani, and J. H. P. Tyman. European Patent App. 80 3007020 (1980).
16. J. H. P. Tyman. J. Chem. Soc. Perkin Trans. I 1639 (1973).

17. J. H. P. Tyman, V. Tychopoulos, and B. A. Colenutt. J. Chromatogr. *213*, 287 (1981).
18. J. C. Touchstone, S. S. Levin, and T. Murawec. In: *Quantitative Thin Layer Chromatography*, J. C. Toughstone (ed.), p. 6, Wiley-Interscience, New York, 1974.
19. O. S. Privett, K. A. Dougherty, and W. L. Erdahl. In: *Quantitative Thin Layer Chromatography*, J. C. Touchstone (ed.) p. 72, Wiley-Interscience, New York, 1974.
20. R. R. Goodall. J. Chromatogr. *123*, 5 (1976).
21. J. H. P. Tyman and M. A. Kashani. Unpublished work.
22. M. Sarsúnová, B. Kakác, and L. Krasnec. J. Chromgatogr. *48*, 253 (1970).
23. J. B. Shute, H. J. de Jong, and H. A. Dingjan. Pharm. Weekbl. *105*, 1025 (1970).
24. E. S. Stern and C. J. Timmons. *Introduction to Electronic Absorption Spectroscopy in Organic Chemistry*, 3d ed., p. 202, Arnold, London, 1970.
25. J. H. P. Tyman, V. Tychopoulos, and P. Chan. J. Chromatogr. *303*, 1984 (1984).
26. J. H. P. Tyman. J. Chromatogr. 156, 255 (1978).
27. J. H. P. Tyman. J. Chromatogr. *111*, 277 (1975).
28. J. H. P. Tyman. J. Chromatogr. *136*, 289 (1977).
29. J. H. P. Tyman and R. A. Stone. Unpublished work.
30. Y. Yamauchi, R. Oshima, and J. Kumanotani. J. Chromatogr. *303*, 137 (1984).
31. J. H. P. Tyman and V. Tychopoulos. Unpublished work.
32. W. O. Ord and P. C. Bamford. Chem. Ind. (London) 1681 (1966); 277 (1967).
33. H. L. Bird, H. F. Brickley, J. P. Corner, P. E. Hartshaw, and M. L. Johnson. Anal. Chem. *35*, 346 (1963).
34. M. A. Millett, W. E. Moore, and J. F. Salman. Anal. Chem. *36*, 491 (1964).
35. H. Ganshirt and K. Morianz. Arch. Pharm. *293*, 1065 (1960).
36. H. Ganshirt and J. Polderman. J. Chromatogr. *16*, 510 (1963).
37. R. Foppiano and B. B. Brown. J. Pharm. Sci. *54*, 206 (1965).
38. H. Ganshirt. Arch. Pharm. *296*, 129 (1963).
39. J. L. McLaughlin, J. E. Gorgan, and A. G. Paul. J. Pharm. Sci. *53*, 306 (1964).
40. J S. Matthews, A. L. Pereda, and A. Aquilera. J. Chromatogr. *9*, 331 (1962).

41. W. Poethke and W. Kinze. Arch. Pharm. *297*, 593 (1964).
42. G. Pataki. Chromatographia *1*, 492 (1968).
43. R. Klaus. J. Chromatogr. *40*, 235 (1969).
44. J. Goldman and R. R. Goodall. J. Chromatogr. *40*, 345 (1969).
45. J. Goldman and R. R. Goodall. J. Chromatogr. *47*, 386 (1970).
46. W. Schlemmer. J. Chromatogr. *63*, 121 (1971).
47. H. J. Petrowitz. In: *6th International Symposium on Chromatography and Electrophoresis*, p. 186, Brussels, 1970, Ann Arbor Sci. Publ., Ann Arbor, Mich., 1971.
48. J. G. Kirchner. J. Chromatogr. *63*, 3, 73 (1971).
49. V. Pollak. J. Chromatogr. *63*, 145 (1971).
50. V. Pollak and A. A. Boulton. J. Chromatogr. *115*, 335, 349 (1975).
51. V. Pollak. J. Chromatogr. *123*, 11 (1976).
52. L. R. Treiber. J. Chromatogr. *123*, 23 (1976).
53. P. Vestergaard, A. Bachman, T. Piti, and M. Kohn. J. Chromatogr. *111*, 75 (1975).
54. R. G. Ackman. Chem. Ind. (London) 715 (1981).

6

Assay of Steroids and Related Derivatives

GÁBOR SZEPESI *Chemical Works of Gedeon Richter, Ltd., Budapest, Hungary*

I. INTRODUCTION

The quantitative application of thin-layer chromatography in steroid analysis is almost as old as the TLC technique itself. A large number of books, chapters, reviews, and fundamental papers have dealt with the theory and practice of the TLC separation of steroids and related compounds; the most important of these were written by Görög (1), Görög and Szász (2), Heftmann (3), Lisboa (4), and Neher (5). These books are recommended as basic literature sources of information.

In this chapter the most important practical considerations in industrial steroid analysis are summarized. Special attention is paid to the quantitative aspects of the methods as well as to the requirements that must be met by TLC methods used to solve particular analytical problems.

Before discussing the use of TLC in quantitative steroid analysis, it must be emphasized that good chromatographic resolution is the most important factor in achieving reliable results. Owing to the existence of different groups of steroids (sex hormones, corticosteroids, cholesterol derivatives, vitamin D and related compounds, bile acids, cardiac glycosides, diosgenin-like compounds, etc.), the polarities of the steroid compounds investigated vary over a wide range from highly polar compounds (e.g., digitalis glycosides) to less polar compounds (vitamin D). There is no single chromatographic system generally applicable for every type of steroids, but some of the recommended methods can be used when related steroids are separated. Special consideration must be given to chromatographic resolution when in situ densitometric evaluation is used for quantitation. For this reason, a separate section in this chapter deals with separation and detection problems in quantitative steroid analysis.

II. MOST IMPORTANT ANALYTICAL TASKS IN INDUSTRIAL STEROID ANALYSIS (CONNECTION BETWEEN METHOD USED AND ANALYTICAL PROBLEM TO BE SOLVED)

A. Determination of Steroids in Plant Extracts and Fermentation Mixtures

The analytical task is to determine some related steroid compounds present at low concentration after their selective separation from each other and from the unknown so-called background materials, which occur at high concentration.

The most important requirements of the quantitative TLC method are the following:

Precise and accurate sample preparation involving optimization of the extraction, cleanup, and sample concentration procedures.
High resolution and selectivity of the TLC system. The spots of the main components should not overlap each other and other compounds.
Sensitive detection, by use of a selective developing reagent or suitable detection wavelength to decrease interference from non-steroidal compounds.
Suitable detection limit for the steroid components. If necessary, preconcentration prior to chromatography or higher sample loading on the plate can be used.
The necessary accuracy and precision.
Optimal quantitative recovery of the steroids from the layer and optimal color reaction (if necessary) for the spectrophotometric measurement, if the spot elution technique is used for quantitation.

In industrial steroid analysis two main applications of this type of TLC separation exist: the determination of cardiac glycosides and sapogenins (6) in plants, and the assay of prednisolone and hydrocortisone in fermentation mixtures (7). The problems that arise in these analyses may be demonstrated by the determination of cardiac glycosides in plants by quantitative TLC.

The success of the assay depends highly on the extraction procedure and the preparation of the drug for extraction. When the leaves are dried, ground, and macerated with the extraction liquid, enzymatic degradation of glycosides can occur, resulting in partial splitting of the terminal glucose unit from the molecules, and the degradation product formed can also be further hydrolyzed. For this reason the sample preparation procedure must be very carefully selected, especially if the genuine composition of cardiac glycosides in the plant sample is determined. Prior to chromatographic separation, pigments, sapogenins, and other impurities must be removed from the extract by addition of lead salts (8–11).

The analysis is simple when the aim is not to obtain data on the contents of all the glycosides, but to determine the digoxin and digitoxin contents of the drug, together with lanatoside C and A as well as acetyldigoxin and acetyldigitoxin, which are transformed to digoxin and digitoxin, respectively (12). The glucose unit can be removed enzymatically by prolonged maceration, and the deacetylation can be carried out with sodium hydroxide or sodium methoxide (13, 14).

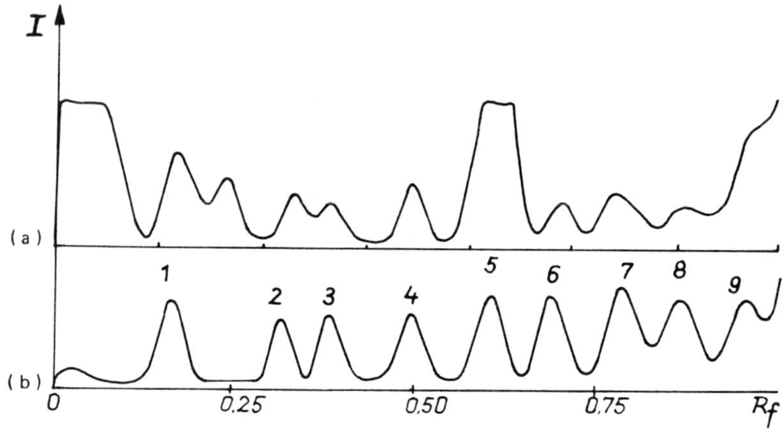

Figure 1 Densitogram of *Digitalis lanata* plant extract and a mixture of standards by the repeated development technique. Chromatoplate: cellulose (E. Merck, Darmstadt, FRG) impregnated with acetone and formamide (10:2). Eluent 1, xylene-methyl ethyl ketone (1:1) saturated with formamide; eluent 2, chloroform; detection, in situ fluorodensitometry (excitation at 366 nm, emission at 436 nm). Compounds: 1, digoxigenin-monodigitoxoside; 2, digoxigenin-bisdigitoxoside; 3, digoxigenin; 4, gitoxigenin-monodigitoxoside; 5, digoxin; 6, gitoxigenin-bisdigitoxoside; 7, gitoxigenin; 8, digoxoside; 9, gitoxin.

The second problem in the analysis is to achieve an acceptable separation of both primary and secondary glycosides. According to data in the literature at least two chromatographic steps must be applied to separate the components (15). Two-dimensional thin-layer chromatography has also been successfully used for the separation of cardenolides of *Digitalis lanata* leaf (16). However, the use of two simple solvent systems should also give good efficiency and a relatively high separation number, as shown in Figure 1 (17). By using overpressurized thin-layer chromatography (OPTLC), much higher separating power can be achieved (17) (a short introduction to this technique will be given in Section III.B). Finally, the application of high-performance thin-layer chromatography to cardiac glycosides results in a great improvement in separation efficiency (17, 18).

The third problem in the analysis is quantitation, due to the low wavelength of the ultraviolet (UV) maximum. A large variety of spray reagents are available for visualization, and some of them are suitable for quantitative purposes. The spot elution technique associated with various colorimetric methods (8, 10, 14, 19–27) can be used only for the determination of a limited number of the components, whereas all the components can be quantified by in situ densitometric or, preferably, fluorimetric scanning methods (17, 28, 29). However—and this is very important—the extinction of separated components after color reaction or the fluorimetric emission after derivatization is highly dependent on the structure of the native glycosides. For this reason, a mixture of reference standards is simultaneously investigated and evaluated.

B. Determination of Bulk Steroids (Intermediates, Crude Products, Active Ingredients)

The main analytical task can be divided into three groups:

Purity control of steroids to check the number and level of impurities present in the steroid sample
Separation and quantitative determination of closely related steroid compounds such as isomers and so-called difficult pairs
Determination of the active ingredient content of bulk steroids

The most chromatographic parameters in the above cases are the selectivity and efficiency of the separation. When purity control of steroids is performed, detectability is also important because the impurities occurring in low concentration must be determined. The quantity of the individual contaminant usually does not allow use of the spot elution technique with a subsequent colorimetric method. In practice, semiquantitative estimation of the spots is carried out by visual comparison with spots for reference materials. Owing to the great differences between the color or fluorescence produced by chemical reaction on the plate, quantitation of different compounds is problematic if no reference material can be used. The use of in situ TLC densitometry is promising, but it does not seem to have been exploited in this field.

Figure 2 shows an example of a satisfactory separation of the impurities occurring in low concentration in chenodeoxycholic acid. In this case a 1-mg sample was applied to the plate and the separated spots were evaluated by densitometry after spraying with a suitable reagent (30).

The accuracy and precision of the methods are not very critical factors in the analysis, and the sample preparation requires precautions only in cases where the substance (or its impurities) can easily decompose or transform in solution.

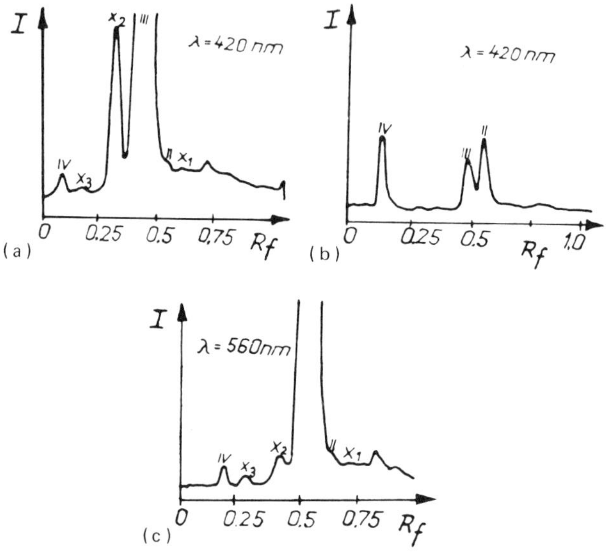

Figure 2 Separation and determination of impurities present in bulk chenodeoxycholic acid. Conditions were as follows. Chromatoplate: silica gel 60 F_{254} (E. Merck, Darmstadt, FRG). Eluent: chloroform-ethyl acetate-glacial acetic acid-2-methoxyethanol (9:9:2:1). Detection: in situ spectrodensitometry at 420 nm (a, b) and 560 nm (c) in reflection mode after development with *p*-toluenesulfonic acid. Compounds: *I*, litocholic acid; *II*, deoxycholic acid; *III*, chenodeoxycholic acid; *IV*, cholic acid.

In almost every group of steroids analytical problems may occur which necessitate the separation of isomers and difficult pairs. In these cases the TLC systems must be very carefully optimized, and sometimes one chromatographic run is not enough to achieve satisfactory separation. A repeated development technique, two-dimensional chromatography, or overrun chromatography can be applied successfully to further increase the chromatographic resolution.

Difficult pairs occur in the pregnane series, and common TLC systems (silica gel with toluene-ethyl acetate or benzene-ethanol eluents) and 2 to 10 successive developments have been used to separate isomeric pairs such as 5α- and 5β-pregnane-3,20-dione, 5α-pregnane-3α,20α-diol and its 20β- isomer, etc. (31, 32). Table 1 shows the conditions of the separation of epimeric pairs of α- and

Table 1 Application of Thin-Layer Chromatography in Industrial Steroid Analysis

Author	Materials separated	Analytical tasks	Chromatographic conditions			
			Layer	Eluent	Detection (reagent)	Determination
			I. Sex hormones			
Amin and Hassenbach (47)	Ethynylestradiol, levonorgestrel	C.U. tests	Silica gel	Cyclohexane-diethyl ether-acetic acid (50:50:1)	Sulfuric acid	In situ UV densitometry
Kammerl and Mutschler (71)	Ethynylestradiol,	C.U. tests	Silica gel	Diethyl ether-petroleum ether (6:4)	Triethanolamine	In situ UV densitometry
Molnár et al. (46)	Ethynylestradiol	Determination in tablets	Silica gel	Benzene-ethyl acetate (6:2); chloroform-hexane-ethyl acetate-methanol (95:20:5:3)	UV	In situ UV densitometry
Sperandeo et al.(72)	Formebolone	Purity test	Silica gel	Ethyl acetate-benzene-acetic acid-water (10:10:4:10)	UV	In situ UV densitometry
Van der Merwe et al. (73)	Spironolactone, cancerone	Assay	Silica gel	Carbon tetrachloride-ethyl acetate (2:3)	Sulfuric acid	In situ fluorescence densitometry

Table 1 (Continued)

Author	Materials separated	Analytical tasks	Chromatographic conditions: Layer	Eluent	Detection (reagent)	Determination
Wintersteiger and Gamse (74)	Ketosteroids	Determination in formulations	Silica gel	Chloroform-benzene-methanol (50:15:5)	Prederivatization dansyl hydrazine	In situ fluorescence densitometry
			II. Corticosteroids			
Bailey et al. (33)	Corticosteroids	Purity test in formulations	Silica gel	Five different solvent mixtures	Sulfuric acid	Semiquantitation
Bishara and Jakovljevic (19)	Fluorandrenolone acetonide	Purity test	Silica gel	Benzene-ethyl acetate	UV	Spot elution, then tetrazolium blue at 520 nm
Huf et al. (44)	Hydrocortisone acetate	Determination in ointment	Silica gel	Benzene-acetone-water (60:40:1)	*p*-Toulenesulfonic acid	In situ UV densitometry
Huf et al. (75)	Hydrocortisone, prednisolone, bectomethasone acetate	Determination in formulations	Silica gel	Cyclohexane-ethyl acetate-water (75:25:1)	Pyrolyzed at 170°C	In situ UV densitometry
Kammerl and Mutschler (71)	Prednisolone	Determination in tablets	Silica gel	Ethyl acetate-cyclohexane-ethanol (4:4:2)	UV	In situ UV densitometry

Lanouette and and Lodge (76)	Corticosteroids	Stability test in formulations	Silica gel	Methylene chloride-dioxane-water	UV	Semiquantitation
Macioci and Fiotek (68)	Hydrocortisone	Determination in ointment	Silica gel	Methylene-chloride-butanol-acetone, (254:30:14)	UV	Spot elution, then UV at 240 nm
Vanderhaeghe and Hoekins (77)	Corticosteroids	Purity test	Silica gel	Five different mobile phases	Sulfuric acid	Semiquantitation
			III. Vitamins D			
Bolliger and König (52)	Vitamins D	Assay in formulations	Silica gel	Cyclohexane-diethyl ether (1:1)	$SbCl_3$	Spot elution, then colorimetry
Galecka et al. (78)	Vitamins D	Purity test	Silica gel	Chloroform	$SbCl_3$	Semiquantitation
Heaysmann and Swayer (79)	Vitamins D	Purity test in formulations	Silica gel	Cyclohexane-diethyl ether (4:1)	Sulfuric acid	Semiquantitation
Hom et al. (80)	Dihydrotachysterol	Purity test in formulations	Silica gel	Dichloroethane	UV	Spot elution, then UV
Janecke and Maas-Goebels (81)	Vitamins D and decomposition products	Purity test	Silica gel	Hexane-ethyl acetate (9:1)	Phosphotungstic acid, $SbCl_3$	Semiquantitation
Kobayashi and Adachi (82)	Vitamins D	Determination in formulations	Silica gel	Hexane-ethyl acetate (4:1)	UV	Spot elution, then GC

Table 1 (Continued)

Author	Materials separated	Analytical tasks	Chromatographic conditions			
			Layer	Eluent	Detection (reagent)	Determination
Koleva et al. (48, 49)	Vitamins D	Determination in formulations	Silica gel	Cyclohexane-diethyl ether (4:1)	Cerium sulfate	In situ UV densitometry
Richter and Ropte (83)	Vitamins D	Assay	Silica gel + PEG	Light petroleum-benzene (1:1)	UV	Spot elution, then $SbCl_3$ colorimetry
Vachek et al. (84)	Vitamins D	Assay in formulations	Alumina	Chloroform	UV	Spot elution, then UV
Wardynska (85)	Vitamins D	Assay in formulations	Silica gel	Cyclohexane-ethyl acetate (1:1)	UV	Spot elution, then $SbCl_3$ colorimetry
IV. Cardiac glycosides						
Banczer et al. (86)	Digitalis glycosides	Determination in plants	Silica gel	Methylene chloride-methanol-acetic acid (90:14:0.5)	—	In situ UV densitometry
Bican-Fister and Merkas (87)	Digitalis glycosides	Assay	Silica gel	Chloroform-methanol (88:12)	Dixanthylurea	Spot elution, then UV
Corona and Roiteri (20)	K-strophantine glycosides	Assay	Silica gel	Chloroform-methanol-acetic acid (85:13:2)	Dixanthylurea	Spot elution, then picric acid, UV

Evans et al. (21)	Digitalis glycosides	Assay in plants and formulations	Silica gel	Ethyl acetate	$SbCl_3$	In situ UV densitometry
Heusser (24)	Digitalis glycosides	Determination in plants	Silica gel	Chloroform-methanol (92:8)	—	Spot elution, then color reaction
Horváth (88)	Digitalis glycosides	Determination in plants	Cellulose + formamide	Xylene-methyl ethyl ketone (1:1), chloroform	Chloramin T	In situ fluorodensitometry
Horváth (14)	Digitalis glycosides	Determination in plants	Silica gel	Methylene chloride-methanol-water (85:13.5:1.5; 80:18:2)	Xanthidrol	In situ spectrodensitometry
Horváth et al. (17)	Digitalis glycosides	Determination in plants	Silica gel	Methylene chloride-methanol-water (90:9:1)	$SbCl_3$	In situ spectrodensitometry
Kartnig and Kobosil (29)	Digitalis glycosides	Determination in plants	Silica gel + formamide	Chloroform-tetrahydrofurane-formamide (5:11:0.5)	—	In situ spectrodensitometry
Khafagy and Girgis (8)	Digitalis glycosides	Determination in plants and formulations	Silica gel	Methylene chloride-methanol-formamide (89.5:9.5:1)	—	Spot elution, then picric acid UV
Mincsovics et al. (18)	Digitalis glycosides	Assay and purity test	Silica gel	Acetonitrile-methyl ethyl ketone (89:5:6)	Phosphoric acid	In situ spectrodensitometry UV

Table 1 (Continued)

Author	Materials separated	Analytical tasks	Chromatographic conditions			
			Layer	Eluent	Detection (reagent)	Determination
V. Diosgenin and related sapogenins						
Blunden and Hardman (12)	Sapogenins	Determination in plants	Silica gel	Hexane-ethyl acetate (12:1)	$SbCl_3$	In situ spectrodensitometry
Blunden et al. (89)	Diosgenin	Determination in plants	Silica gel	Hexane-ethyl acetate (4:1)	$SbCl_3$	In situ spectrodensitometry
Brain and Hardman (90)	Sapogenins	Determination in plants	Silica gel	Hexane-acetone (8:1)	$SbCl_3$	In situ spectrodensitometry
Csizér and Görög (91)	Diosgenin	Assay	Silica gel	Methylene chloride-methanol (97:3)	—	Spot elution, then titration with bromine
Lockwood et al. (92)	Sterols and sapogenins	Determination in plants	Silica gel	Hexane-acetone (4:1)	$SbCl_3$	In situ spectrodensitometry
Lope-Sanchez et al. (93)	Diosgenin	Determination in plants	Silica gel	Chloroform-acetone (8:2)	—	Spot elution, then sulfuric acid UV
Nikolov et al. (94)	Ruscogenin	Determination in plants	Silica gel	Cyclohexane-ethyl acetate (1:1)	*p*-Dimethylaminobenzaldehyde	In situ spectrodensitometry

Pkheidze and Kemertalidze (95)	Tigogenin	Purity test	Silica gel	Chloroform-ethanol (25:1)	UV + iodine	Semiquantitation
			VI. Miscellaneous steroids			
Hardman and Benjamin (96)	Phytoecdysones	Determination in plants	Silica gel	Methylene chloride-ethanol (5:1)	UV	Spot elution, then UV
Kartnig and Mikula (38)	Sterols	Purity test	MgO	Cyclohexane-diethyl ether-acetic acid	—	Spot elution, then sulfuric acid UV
Kinger and Michoel (97)	Pancuronium bromide	Stability test in injection	Silica gel	Butanol-pyridine-acetic acid-aqueous ammonium chloride (60:40:12:48)	Dragendorf	In situ spectrodensitometry
Kovács et al. (98)	Pipecuronium bromide	Stability test in injection	Alumina	Isobutanol-pyridine-acetic acid-aqueous ammonium chloride (90:44:13.2:528)	Iodine	In situ spectrodensitometry
Szepesi et al. (30)	Chenodeoxycholic acid	Purity test	Silica gel	Chloroform-ethyl acetate-acetic acid-2-methoxyethanol (9:9:2:1)	Sulfuric acid	In situ spectrodensitometry

β-dihydroequilin, sterols, vitamin D compounds, digitalis glycosides, α- and β-ecdysones, and sapogenins (33–42). The selectivity and efficiency of the isomer separations can also be increased by using the OPTLC technique (30, 43).

When the analytical task is determing the active ingredient content of bulk steroids, the selectivity and efficiency of the separation are not as critical factors in the analysis as in the other two cases mentioned. The aim is to separate the main components from the impurities, but separation of the impurities from each other is not required. The accuracy and precision of the method are much more important. A few years ago, quantitative TLC with spot elution was a standard procedure in the U.S. Pharmacopoeia (USP) for the assay of different steroids (44, 45). This technique, however, requires special skills and expertise to remain within acceptable limits of error because of the deviation of sample application, the inaccuracy in the location of the spots and in scraping them off, and the losses due to irreversible adsorption or decomposition during chromatography, elution, etc.

Some of the sources of error can be eliminated by simultaneous investigation of steroid reference standards and by using the data pair technique for quantitative evaluation. However, the assay method for bulk steroids has been changed in several cases in the latest editions of USP XX. (supplements and addendums). Instead of quantitative TLC methods, high-performance liquid chromatography (HPLC) is used for the assay of some steroids (45).

C. Assay of Steroids in Pharmaceutical Formulations

1. *Determination of Active Ingredient Content of Single or Multicomponent Steroid Preparations*

The main source of analytical problems is related to quantitative recovery of the active substances. Steroids are used therapeutically in practically every kind of pharmaceutical formulation, including tablets, oily or aqueous injections, capsules, suppositories, and ointments, and very frequently combinations are administered in which the steroids occur in very different concentrations.

In most cases, separation of the active ingredients from each other and from the excipients is not a serious problem because of the different nature of the steroids used in the formulation and the selectivity of the color reaction used for the quantitation. Problems arise when side reactions occur during extraction or elution resulting in several by-products derived from active ingredi-

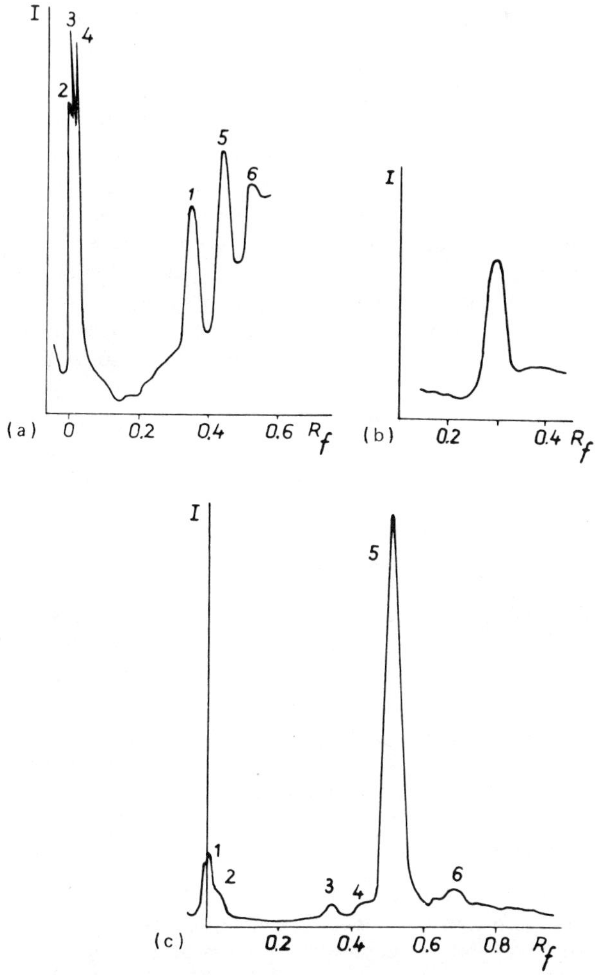

Figure 3 Determination of ethynylestradiol in contraceptive pills. (a) Chromatogram of contraceptive pills extracted with aqueous hydrochloric acid; (b) chromatogram of ethynylestradiol after enzymatic degradation of starch; (c) chromatogram of methyltestosterone and ethynylestradiol-containing tablets. Conditions were as follows. Chromatoplate: silica gel 60 (E. Merck, Darmstadt, FRG), Mobile phases: (a) chloroform-ethyl acetate (8:2); (b) chloroform-hexane-ethyl acetate-methanol (95:20:5:3); (c) hexane-ethyl acetate-diethyl amine (5:3:2). Evaluation: in situ spectrodensityometry in reflection mode at 280 nm. Compounds: ethynylestradiol (<u>1</u> in a and <u>3</u> in c), methyltestosterone (<u>5</u> in c).

ents occurs in very low concentration compared to the others, and when a small amount of impurities present in the active substance used in higher concentration interferes with the separation. This is illustrated in Figure 3, which shows the determination of ethynylestradiol in contraceptive pills (46). When a mixture of methanol and aqueous hydrochloric acid is used to extract the active ingredients (this procedure is generally accepted in the determination of ethynylestradiol in tablets), several by-products are formed. These by-products probably do not disturb the determination based on the color reaction, but they can cause serious problem in the TLC separation (Figure 3a). If the starch is subjected to enzymatic degradation with α-amylase, extraction of ethynylestradiol from the tablets is simple and no by-product is formed during the sample preparation step (Figures 3b and 3c).

Regarding quantitation of the separated steroids, the popularity of methods based on spot elution has decreased considerably, although such methods are still used when the eluted spots are quantified by spectrophotometry, colorimetry, fluorimetry, etc. Recent publications indicate the growing popularity of in situ densitometric methods. Direct densitometry in the reflectance mode has not been used often (see the example of ethynylestradiol in Figure 3). The assay employing a suitable reaction on the plate is more frequently applied. As in the example in Figure 3, the ethynylestradiol content of contraceptive pills can be determined after TLC separation by spraying with sufuric acid-methanol reagent and heating at 120°C, followed by measuring the substance in the reflectance mode at 530 nm or in the fluoresence mode using 365 nm for excitation and 560 nm for emission (47).

2. *Stability Assay of Steroid Preparations*

Besides the extraction problems mentioned above, the most critical factors in the determination are the selectivity, efficiency, and sensitivity of the assay method. Decomposition products of steroids occurring in low concentration must be separated and determined in the presence of active ingredients. The impurities originate from production and different excipients. When the decomposition products are determined, the accuracy and precision of the method are not very problematic; a relative standard deviation of about ±5% should be acceptable. In determining the active ingredient content, however, this value must be much lower. For this reason, quantitation by in situ densitometry seems to be the only acceptable method for the stability assay of steroid preparations. An example is the determination of vitamins D_2 and D_3 in oily and aqueous injections in the presence of their decomposition products (48, 49).

Similarly, the stability of dehydroepiandrosterone enanthate tablets was tested by in situ densitometry after spraying with sulfuric acid-methanol reagent (50). It should be noted that due to the complexity of the analytical requirements, other chromatographic techniques (primarily HPLC) have superseded TLC methods in this field.

3. *Purity Tests of Steorid Formulated Products*

Sometimes the semiquantitative estimation of impurities and decomposition products in steroid formulated products is sufficient to control the quality of the preparation. In such a case the most important thing is to find the chromatographic conditions that provide the required resolution. Table 1 shows a few examples of this type of separation. For instance, the quality of the formulated products of cardiac glycosides was investigated by quantitative TLC, using the spot elution technique and color reaction (8).

4. *Content Uniformity Tests of Steroid-Containing Tablets and Capsules*

Because most oral contraceptives contain more than one active ingredient, and in many cases one of these ingredients occurs in low concentration, the uniformity of the steroid content is very important. Quantitative TLC accomplished by in situ densitometry is a very suitable technique for determining the uniformity of the steroid content. In most cases, separation of the active ingredients is not very difficult; neither is the demand on the accuracy and precision of the method very high. The sample preparation, however, requires consideration of the principles mentioned in connection with the assay of steroid formulations. The spot elution technique, which is usually time and labor consuming, is not recommended for this type of analysis.

Table 1 shows some examples of content uniformity tests for steroids-containing tablets, such as the direct quantitative TLC determination of levonorgestrel and ethynylestradiol in oral contraceptives by diffuse reflectance and fluorescence methods (47). The most important quantitative TLC systems used for the separation and quantitative determinations of steroids are also shown in Table 1.

III. GENERAL CONSIDERATIONS AND PROBLEMS IN QUANTITATIVE TLC DETERMINATIONS OF STEROIDS

This section discusses practical problems that must be considered when steroids are investigated and assayed.

A. Stability of Steroids During Analytical Procedures

Although steroids are considered stable compounds, they can decompose. One source of decomposition may be the sample preparation method, as mentioned in the previous sections. Another source, not mentioned previously, is the oxidative decomposition of steroids adsorbed on the surface of a polar adsorbent such as silica gel. In this form, some steroids are sensitive to atmospheric oxygen, which may cause problems especially in separation and analysis in the low concentration range. The decomposition can be suppressed by the application of antioxidants, e.g., by using ascorbic acid in the eluent or by pretreating the sorbent with an ethanolic solution of ascorbic acid (51).

The D vitamins are the most sensitive compounds among the steroids. They are sensitive to atmospheric oxygen, traces of acids and bases, light, heat, etc. To avoid decomposition and to obtain reliable, reproducible results, the following precautions are recommended (52).

All experiments should be carried out in a cool place, protected from exposure to direct daylight.

Chromatography should be carried out under nitrogen gas or with an antioxidant dissolved in the eluent.

Highly purified solvents should be used.

Since the D vitamins decompose most easily on the surface of silica gel, the sample should not be dried during the spotting and the chromatoplate should be placed in the chamber immediately after sample application. After the run the spots should be located, scraped off, and eluted in the shortest time possible. The duration of UV irradiation at the location should be minimal.

B. Separation Problems

For the separation of steroids, silica gel layers with a variety of eluent systems have been used most widely. As previously mentioned, because of the very different polarities of the steroids investigated no generally applicable chromatography system can be recommended. In Refs. 1 to 5 the TLC systems most suitable for solving the majority of separation problems within an individual group of steroids (sex hormones, corticosteroids, D vitamins, cholesterols, bile acids, cardiac glycosides, etc.) can be found. Some delicate problems such as the separation of difficult pairs of steroids have been mentioned. This section will discuss some methods for improving the selectivity and/or efficiency of separations.

When the silica gel alone is not a sufficiently good adsorbent for solving the separation problems, the results can be improved by using the following approaches.

1. A multiple-development technique, two-dimensional chromatography, and overrun chromatogrpahy used on silica.

2. Impregnation of the silica gel with solvents or reagents. For example, this should be used for the separation of cholesterol and its related derivatives. As sterols are the least polar group of steroids, their separation from other groups of steroids and from other lipids is relatively easy. Many systems described in the literature are suitable for this purpose. The separation of structurally closely related sterols is much more difficult. On silica gel only poor separation efficiency can be achieved. The chromatographic resolution can be improved if silica gel containing silver nitrate is used for the separation. In this case the formation of π-complexes between silver ions and the double bonds of the unsaturated sterols results in satisfactorily separated spots (39, 53–55). Similarly, good results can be achieved when the silica gel layer is impregnated with paraffin or lower hydrocarbons (54). In the case of cardiac glycosides, impregnation of the silica layer with boric acid or formamide is helpful in solving delicate separation problems (56, 57).

3. Chromatography on an adsorbent other than silica gel. Alumina is also a good stationary phase for the separation of sex hormones, corticosteroids, D vitamins, etc. The popularity of reversed phase layers in steroid analysis is not as great as in other fields of pharmaceutical analysis. However, in recent years some interesting applications have been published, and it is worth mentioning that USP XX prescribes the use of a chemically bonded octadecyl silica layer for purity testing of digoxin, using semiquantitative estimation for the quantitation (45d). In special cases polyamide layers (58), impregnated cellulose (59), polyvinylpyrrolidone layers (60), and magnesium oxide (38, 39) have been successfully applied.

4. Overpressurized thin-layer chromatography (OPTLC). Further improvements in the separation of steroids can be expected with the OPLC technique, which was introduced a few years ago by Tyihák and co-workers (61, 52). The technique is based on a concept similar to that underlying liquid chromatography. The solvents are passed through an ultramicro chamber by an appropriate pump system, and the sorbent layer is completely covered in the chamber with a flexible membrane under external pressure.

In this manner the vapor space over the layer can be excluded, resulting in more reproducible retention, better resolution, and faster development than in classical TLC systems. Linear migration of the solvent front in the ultramicrochamber is ensured by impregnating the sides of the layer and by placing a narrow plastic sheet on the layer behind the starting line.

The theoretical background of the OPLC techniques can be summarized as follows. In classical TLC the linear flow velocity can be expressed by the following equation:

$$k^{TLC} = \frac{x^2}{t}$$

where x is the distance of eluent migration (running distance) and t the time required to cover the distance x. This equation shows that the linear flow velocity decreases with increasing running distance, which can cause a loss in separation efficiency. In OPTLC, the linear flow velocity is constant and can be expressed by the following equation:

$$k^{OPTLC} = \frac{x}{t}$$

In this case separation efficiency is practically independent of the running distance on high-performance layers with different developelution. Figure 4 shows the dependence of the HETP values on the running distance on high-performance layers with different developing techniques.

In Figure 5 the separation of digitalis glycosides in a generally used standard chamber is compared to that of development in a pressurized ultramicro chamber (PUM chamber) with silica gel TLC and HPTLC plates (17). It can be seen that the chromatographic resolution may be increased considerably by using a PUM chamber and an HPTLC plate. In this case the optimum migration distance is much longer, and the HETP value is smaller than in the usual HPTLC investigations. Figure 6 shows the dependence of migration distance of the solvent front on the developing time in different chamber systems. The migration time of the solvent front is much smaller in the PUM chamber, so the higher separation number in OPTLC is a consequence of the decrease in migration time available for spreading the spots.

The most important features of the OPTLC technique can be summarized as follows:

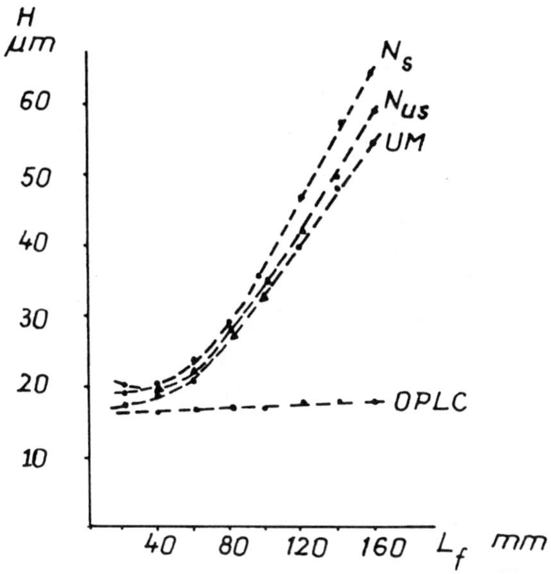

Figure 4 Dependence of HETP values on the running distance on high-performance layers with different developing techniques. N_s, normal saturated chamber; N_{us}, normal unsaturated chamber; UM, ultramicro chamber; OPTLC, pressurized ultramicro chamber.

The method is suitable for fast, efficient, and simultaneous investigation of a large number of samples.

The chromatogram can be developed at a higher migration distance in a reasonable running time on microparticulate stationary phases.

Solvents of high viscosity can also be used for eluent preparation.

The overrunning can be carried out under controlled conditions.

A small amount of eluent is required.

The higher efifciency and better chromatographic resolution provide a more advantageous spot shape for densitometric evaluation.

The HPLC conditions can be modeled.

C. Detection Problems

Although some steroids have a UV absorbance suitable for their detection, and localization of the spots can be based on the appear-

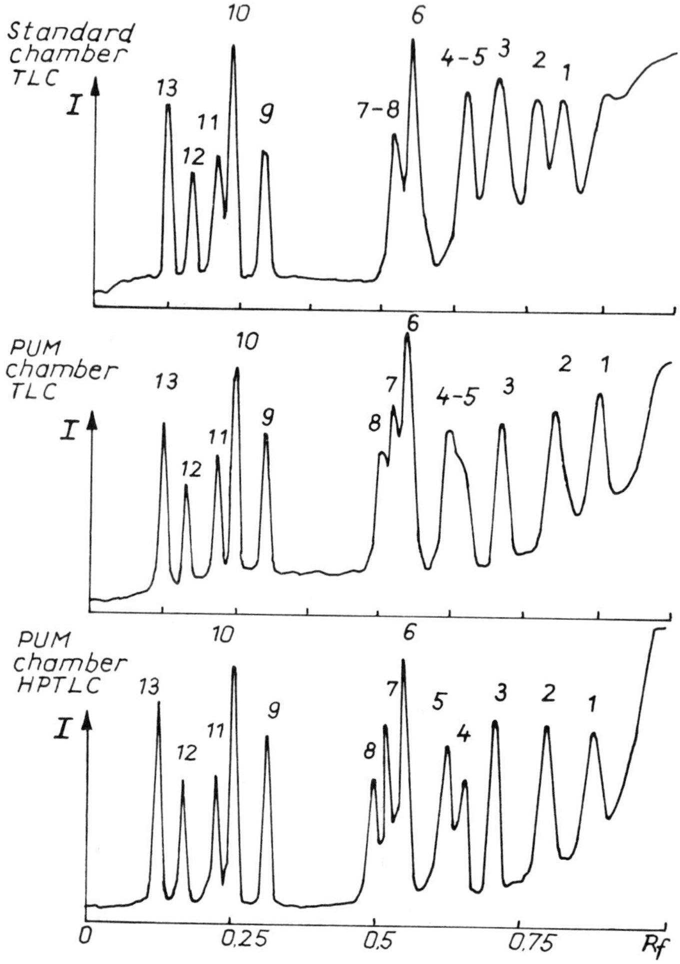

Figure 5 Densitograms of some digitalis glycosides obtained in conventional and PUM chambers with silica gel TLC and HPTLC plates. Mobile phase: acetonitrile-methyl ethyl ketone-water (89:5:6). Conditions for OPTLC: membrane pressure, 1.2 MPa; flow rate, 43.8 cm^3/h. Detection: after fluorogenic reaction, in situ fluorodensitometry (excitation at 366 nm, emission at 436 nm). Compounds: 1, β-acetyldigitoxin; 2, α-acetyldigitoxin; 3, β-acetyldigoxin; 4, α-acetyldigoxin; 5, digitoxin; 6, gitoxin; 7, digoxin; 8, D(+)-digitoxose; 9, lanatoside A; 10, lanatoside B; 11, lanatoside C; 12, deacetyllanatoside A; 13, deacetyllanatoside C.

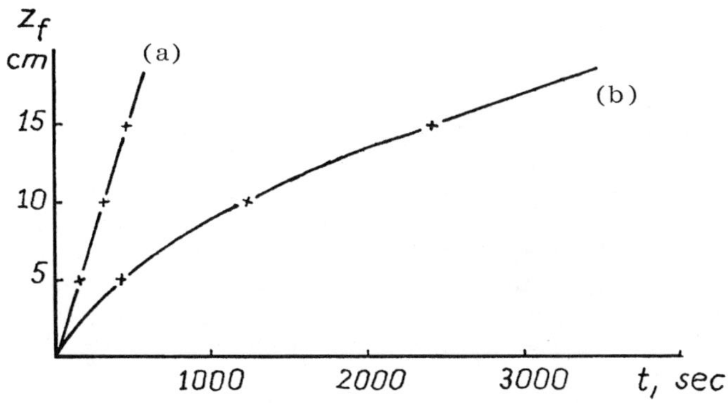

Figure 6 Dependence of running distance on time of development in different chamber systems. (a) PUM chamber; (b) standard chamber.

ance of dark spots in 254-nm UV light if a fluorescent layer is used, most applications still involve methods based on chemical reactions. A great variety of spray reagents for the visualization of steroid spots have been described. However, special attention is needed to stabilize the produced color or the induced fluorescence on the plate. Sulfuric acid-methanol seems to be the most applicable reagent. In addition, phosphoric acid, hydrochloric acid, antimony(III) chloride, etc. can generally be used to detect steroids on the plate or to determine steroids after spot elution.

Another principle of detection should be utilized for quantitation to induce fluorescence on the plate. The simplest way is to heat the plate for several hours at about 120°C in an ammonia atmosphere prepared by in situ heating of ammonium hydrogen carbonate. Although the mechanism of this reaction is unknown, it is considered a simple, generally applicable, and reproducible detection procedure for steroids (63).

Besides derivatization after the separation of steroids, methods employing derivatives prepared prior to application to the plate have some advantages. The most important one is that errors originating from incomplete distribution of the spray reagent can be eliminated (67). In many cases the stability of the steroids investigated or their chromatographic behavior can also be improved by prior derivatization.

A prior derivatization method has been used for separation and detection of keto steroids in 2,4-dinitrophenylhydrazone form (64–66) and of hydrocortisone in the form of the dansyl derivative (6).

D. Problems of Quantitation with the Spot Elution Technique

Because of its simplicity, the spot elution technique is a very popular procedure in the quantitative TLC of steroids. After elution from the TLC plates, the separated steroid compounds can be quantified by colorimetric, fluorimetric, gas chromatographic, mass spectrometric, radiochemical, and other methods. By adding the high efficiency and selectivity of thin-layer chromatography to the sensitivity of the method chosen for quantitation, many very difficult analytical problems can be solved in small laboratories equipped with a basic TLC setup and a spectrophotometer or fluorimeter.

The main problems of the spot elution technique are as follows.

Reliable results can be obtained only with very careful work. Use of appropriate blanks containing adsorbent scraped from an empty part of the chromatogram is required.

Simultaneous use of an appropriate reference standard on the same chromatoplate is necessary.

The accurate location of the spots for subsequent scraping off is not always obvious.

Losses originating from irreversible adsorption or decomposition during chromatography or elution should be minimized.

Nearly quantitative extractions of the spot is required.

The chemical reactions used for quantitation can be optimized in detail. The color or induced fluorescence of the derivatives must be stable for a reasonable time. Only weak formation of chromophores must be obtained with the blank, etc.

The method must be validated. The validation data should be in accordance with the requirements.

As a result of the above factors and of the development of other highly selective chromatographic methods including in situ TLC techniques, the importance of spot elution techniques will probably diminish in the future.

E. Problems of Quantitation with the in situ TLC Technique

Direct analytical methods requiring no spot elution of steroids from the chromatoplate have been used in steroid analysis for a long time. The spot intensities can often be established by visual comparison with reference spots, or from the concentration of a component calculated from the minimum volume of test solution that must be applied on the layer to yield a detectable spot (67). This semiquantitative estimation of impurities is generally used in the different pharmacopoeias (45b).

The in situ quantitation of steroids on the chromatoplate is a much more interesting field. Besides the well-known advantages of quantitative TLC with in situ densitometric methods (the chromatographic separation is independent of the detector in time and space; the chromatogram can be preserved after quantitation, the whole sample can be spotted on the plate; the chromatogram is complete; every component can be detected; fresh adsorbent can be used for the separation of every sample, etc.), the following important features can be added compared to spot elution techniques:

The sensitivity of detection is higher by about 2 orders of magnitude.

Because the samples are small, quantities requiring linear calibration plots can easily be obtained.

Reflectance spectra of the spots can be taken, permitting identification of certain unknown components and making it possible to increase the selectivity by changing the detection wavelength.

Various options exist for the in situ scanning of thin-layer chromatograms of steroids. First, the plate can be scanned in the reflectance mode. Even in the case of relatively poorly UV absorbing steroids, satisfactory detection limits can be achieved at short-wavelength UV maxima. Thus 50 ng of cardiac glycosides can be determined by measuring at a wavelength of 225 nm (38). Second, in the case of UV-active steroids (e.g., corticosteroids), the chromatogram can be scanned on the basis of the fluorescence quenching at 254 nm by using F_{254} silica gel plates (68). Two other possibilities are related to methods based on chemical reactions. These methods are more generally used for scanning chromatograms in the reflectance or in the absorbance mode. In the visible spectral range simple and inexpensive instruments with filter photometers can also be used.

Direct fluorimetric evaluation of the chromatograms after treatment with a suitable fluorogenic reagent is the most sensitive method. As previously mentioned, three different varieties of fluorimetric evaluation can be performed; these involve (a) producing the fluorimetrically active product on the plate by spraying it with a suitable reagent. (b) inducing the fluorescence on the plate by means of heat treatment and/or vaporized reagent, and (c) preparing the fluorescent derivatives prior to the TLC separation.

A new technique for in situ quantitation involves separation on silica gel-coated quartz rods. The quantitation is performed directly by using a flame ionization detector. This elegant technique was introduced by Japanese workers (15, 69), and it has been used for the determination of conjugated bile acids and cardiac glycosides (15, 69, 70). This method is discussed in detail in Chapter 4 of this book.

The problems of in situ quantitation can be summarized as follows.

Simultaneous running of an appropriate reference standard is required. No area normalization method can be used for quantitation.

Errors originate from incomplete distribution of the spray reagent.

The variation of layer thickness along the plate influences the results.

The stability of the product formed by chemical reaction is more or less limited.

The in situ techniques are naturally less labor-intensive than methods based on spot elution. Their high efficiency, selectivity, and sensitivity make these techniques comparable to gas chromatography and high-performance liquid chromatography. The advantages of the in situ TLC techniques compared to GLC and HPLC are that usually no derivatization is necessary, any type of chemical reagent can be used to increase the selectivity and sensitivity of the steroid determination, compounds differing widely in polarity can be separated and determined, scanning the chromatograms requires less time than in GLC and HPLC, and a single densitometer can be used for quantitative evaluation of the chromatograms from several methods.

The main disadvantage of these techniques is their lower accuracy and precision, caused predominantly by the imperfections of the sample application procedure and the characteristics of the sorbent layer.

REFERENCES

1. S. Görög. *Quantitative Analysis of Steroids*, Akadémiai Kiadó, Budapest, Elsevier, Amsterdam, 1983.
2. S. Görög and Gy. Szász. *Analysis of Steroid Hormone Drugs*, Akadémiai Kiadó, Budapest, Elsevier, Amsterdam, 1978.

3. E. Heftmann. *Chromatography of Steroids*, Elsevier, Amsterdam, 1976.

4. B. P. Lisboa. Steroids. In: *Pharmaceutical Applications of Thin-Layer and Paper Chromatography*, K. Macek (ed.), Elsevier, Amsterdam, 1972.

5. R. Neher. TLC of Steroids and Related Compounds. In: *Thin-Layer Chromatography*, E. Stahl (ed.), Springer Verlag, Berlin, 1969.

6. F. Kreuzig. Z. Anal. Chem. *281*, 135 (1976).

7. L. V. Andreev and K. A. Koschenko. Zh. Anal. Khim. *31*, 343 (1976).

8. S. M. Khafagy and A. N. Girgis. Planta Med. *25*, 350 (1974).

9. Ch. B. Lugt. Pharm. Weekbl. *111*, 405 (1978).

10. I. Szilágyi, I. Zámbó, and J. Török. Planta Med. *32*, 60 (1977).

11. M. Wichtl and R. Summer-Irsigler. Planta Med. *7*, 129 (1959).

12. G. Blunden and R. Hardman. J. Chromatogr. *34*, 507 (1968).

13. P. H. Cobb. Analyst *101*, 768 (1976).

14. P. Horváth. Acta Pharm. Hung. *52*, 133 (1982).

15. A. Uji. Yakugaku Zasshi *95*, 214 (1975).

16. C. J. Clarke and P. H. Cobb. J. Chromatogr. *168*, 541 (1979).

17. P. Horváth, G. Szepesi, M. Hoznek, Z. Végh, and E. Mincsovics. In: *Proceedings of the 2nd International Symposium on Instrumental High Performance TLC in Kaiser*, R. E. (ed.), p. 295, Interlaken, 1982.

18. E. Mincsovics, T. J. Székely, M. Hoznek, Z. Végh, I. Zámbó, G. Szepesi, and E. Tyihák. In: *Advances in Steroid Analysis*, S. Görög (ed.), p. 427, Akadémiai Kiadó, Budapest, Elsevier, Amsterdam, 1982.

19. R. H. Bishara and I. M. Jakovljevic. J. Pharm. Sci. *59*, 124 (1970).

20. G. L. Corona and M. Raiteri. J. Chromatogr. *19*, 435 (1965).

21. J. Evans, P. A. Flemous, C. F. Duignan, and P. S. Cowley. J. Chromatogr. *88*, 341 (1974).

22. B. Görlich: Planta Med. *9*, 442 (1961).

23. W. Hauser, T. Kartnig, and G. Verdino. Sci. Pharm. *37*, 149 (1969).

24. D. Heusser. Planta Med. *12*, 237 (1964).

25. L. Hörhammer and H. Wagner. Dtsch. Apoth. Ztg. *107*, 1793 (1967).

26. H. Pötter. Pharmazie *20*, 737 (1965).

27. E. Tschesche and G. Grimmer. Berichte *93*, 1477 (1960).

28. F. Doelker, J. Kapetanidis, and A. Mirimanoff. Pharma. Acta Helv. *44*, 647 (1969).

29. T. Kartnig and P. Kobosil. J. Chromatogr. *138*, 238 (1977).

30. G. Szepesi, K. Dudás, A. Pap, Z. Végh, E. Mincsovics, and E. Tyihák. J. Chromatogr. *237*, 137 (1982).

31. G. Cavina, G. Moretti, and M. Petrella. J. Chromatogr. *103*, 368 (1975).

32. B. P. Lisboa, H. T. Schneider, and H. Breuer. Acta Endocrinol. *82*, 57 (1976).

33. K. Bailey, A. W. By, and B. A. Lodge, J. Chromatogr. *166*, 299 (1978).

34. G. Carignan, M. Lanouette, and B. A. Lodge. J. Chromatogr. *135*, 523 (1977).

35. G. Carignan, M. Lanouette, and B. A. Lodge, J. Can. Pharm. Sci. *12*, 106 (1977).

36. A. M. Dawidar, A. A. Saleh, and S. L. Elmotei. Planta Med. *24*, 367 (1974).

37. M. B. E. Fayez and A. S. Radwan. Z. Anal. Chem. *227*, 30 (1967).

38. T. Kartnig and G. Mikula. J. Chromatogr. *53*, 537 (1970).

39. B. A. Knights. Anal. Biochem. *80*, 324 (1977).

40. F. Mermet-Bouvier. J. Chromatogr. *59*, 226 (1971).

41. M. F. Ruh and C. Black. J. Chromatogr. *116*, 480 (1976).

42. J. Schreiber, O. Aurich, and G. Osske. J. Chromatogr. *12*, 63 (1963).

43. K. Dudás, G. Szepesi, A. Pap, T. Fehér, E. Mincsovics, and E. Tyihák. In: *Advances in Steroid Analysis*, S. Görög (ed.), p. 417, Akademiai Kiadó, Budapest, Elsevier, Amsterdam, 1982.

44. F. A. Huf, J. C. M. Herberts, and M. A. Span. Pharm. Weekbl. *114*, 660 (1979).

45. *United States Pharmacopoeia* XX, USP Convention Inc., Rockville, Md., 1980. (a) pp. 82, 172, 174, 270, 295, 329, 482, 523, 541, 542, 573, 587, 652, 654, 666, 772, 814 (1980); Supplements: pp. 97, 106, 191. (b) pp. 82, 298, 325, 493, 555, 557, 560 (1980); Supplements: pp. 106, 130, 496, 522. (c) pp. 80, 83, 85, 87, 176, 208, 313, 525, 631, 654 815 (1980); Supplements: pp. 89, 97, 99, 105, 132, 133, 144, 146, 147, 148, 191, 254, 274, 506, 514, 522, 524, 690, 722, 795, 877, 878. (d) Supplement: pp. 106 (1981).

46. J. Molnár, M. Gazdag, and G. Szepesi. Pharmazie *37*, 12 (1982).

47. M. Amin and M. Hassenbach. Analyst *104*, 407 (1979).

48. M. Koleva, M. Jonedini, and O. Budewski. Pharmazie *28*, 317 (1973).

49. M. Koleva, M. Jonedini, and O. Budewski. Pharmazie *30*, 168 (1975).

50. M. Amin, A. Seeger, K. Prezewowsky, and P. Fuchs. Pharm. Acta Helv. *49*, 117 (1974).

51. S. Fragcic and Z. Kniewald. J. Chromatogr. *94*, 291 (1974).

52. H. R. Bollinger and A. König. Z. Anal. Chem. *214*, 1 (1965).

53. J. R. Claude. J. Chromatogr. *17*, 596 (1965).

54. J. W. Copius-Peereboom and H. W. Beekes. J. Chromatogr. *17*, 99 (1965).

55. R. Ikan and M. Cudzinovski. J. Chromatogr. *18*, 422 (1965).

56. H. Pötter and H. Bärisch. Pharmazie *27*, 315 (1972).

57. Z. Szeleczky. J. Chromatogr. *178*, 453 (1979).

58. M. S. Scandrett and E. Ross. J. Clin. Chim. Acta *72*, 165 (1976).

59. W. Tinschert and L. Träger. J. Chromatogr. *152*, 447 (1978).

60. J. M. Marcos, J. Moreno, and J. M. Pla Delfina. Cienc. Ind. Pharm. *9*, 92 (1977).

61. E. Mincsovics, H. Kalász, and E. Tyihák. J. Chromatogr. *181*, 293 (1980).

62. E. Tyihák, E. Mincsovics, and H. Kalász. J. Chromatogr. *174*, 75 (1979).

63. R. Segura and A. Gotto. J. Chromatogr. *99*, 643 (1974).

64. L. R. Treiber. J. Chromatogr. *123*, 23 (1976).

65. P. Knapstein, L. R. Treiber, and J. C. Touschstone. Steroids *11*, 915 (1968).

66. R. M. Thompson. J. Steroid Biochem. *7*, 845 (1976).

67. R. E. Graham and C. T. Kenner. J. Pharm. Sci. *62*, 103 (1973).

68. F. Macioci and C. Fiotek. Boll. Chim. Farm. *144*, 468 (1975).

69. T. Okumura, T. Kadono, and A. Iso'o. J. Chromatogr. *108*, 329 (1975).

70. R. Beke, G. A. De Weerdt, and F. Barbier. J. Chromatogr. *193*, 504 (1980).

71. E. Kammerl and E. Mutschler. Pharm. Acta Helv. *50*, 269 (1975).

72. F. Sperandeo, S. Pedrazzini, G. De Marchi, and E. Marre. Boll. Chim. Farm. *118*, 200 (1979).

73. P. J. Van der Merwe, D. G. Müller, and E. C. Clark. J. Chromatogr. *171*, 579 (1979).

74. R. Wintersteiger and E. Gamse. In: *Advances in Steroid Analysis*, S. Görög (ed.), p. 453, Akadémiai Kiadó, Budapest, Elsevier, Amsterdam, 1982.

75. F. A. Huf, M. A. Span, and H. J. De Jong. In: *Advances in Steroid Analysis*, S. Görög (ed.), P. 465, Akadémiai Kiadó, Budapest, Elsevier, Amsterdam, 1982.

76. M. Lanouette and B. A. Lodge. J. Chromatogr. *129*, 475 (1976).

77. H. Vanderhaeghe and J. Hoebius. J. Pharm. Belg. *31*, 25 (1976).

78. H. Galecka, K. Paczek, and H. Wardynska. Acta Pol. Pharm. *29*, 257 (1972).

79. L. T. Heaysmann and E. R. Swayer. Analyst *89*, 529 (1964).

80. F. S. Hom, S. A. Veresh, and W. R. Ebert. J. Assoc. Off. Anal. Chem. *60*, 48 (1977).

81. H. Janecke and I. Maas-Goebels. Z. Anal. Chem. *178*, 161 (1960).

82. T. Kobayashi and A. Adachi. J. Nutr. Sci. Vitaminol. *22*, 41 (1976).

83. J. Richter and D. Ropte. Pharmazie *24*, 601 (1969).

84. J. Vachek, B. Kakac, and M. Sarsunova. Cesk. Farm. *16*, 229 (1967).

85. H. Wardynska. Acta Pol. Pharm. *26*, 63 (1969).

86. E. Banczer, T. Prey, and F. Wurst. Planta Med. *29*, 393 (1976).

87. T. Bican-Fister and J. Merkas. J. Chromatogr. *41*, 91 (1969).

88. P. Horváth. In: *Advances in Steroid Analysis*. S. Görög (ed.), p. 457, Akadémiai Kiadó, Budapest, Elsevier, Amsterdam, 1982.

89. G. Blunden, R. Hardman, and J. C. Morrison. J. Pharm. Sci. *56*, 948 (1967).

90. K. R. Brain and R. Hardman. J. Chromatogr. *38*, 355 (1968).

91. É. Csizér and S. Görög. J. Chromatogr. *76*, 502 (1973).

92. G. B. Lockwood, K. R. Brain, and T. D. Turner, J. Chromatogr. *95*, 250 (1974).

93. G. Lope-Sanchez, J. C. Medina-Acevedo, and R. R. Soto. Analyst *97*, 973 (1972).

94. St. Nikolov, M. Joneidi, and D. Panova. Pharmazie *31*, 611 (1976).

95. T. A. Pkheidze and E. P. Kemertalidze. Khim. Farm. Zh. *10*, 127 (1976).

96. R. Hardman and T. V. Benjamin. J. Chromatogr. *131*, 468 (1977).

97. R. Kinger and A. Michoel. J. Chromatogr. *120*, 234 (1976).

98. P. Kovács, G. Takácsi Nagy, and P. Horváth. Acta Pharm. Hung. *53*, 56, 64, 71 (1983).

7

Quantitative Thin-Layer Chromatography of Vitamins

TOM R. WATKINS *Hunter College, The City University of New York, New York, New York*

I. INTRODUCTION

Labeling laws for foods, pharmaceuticals, and cosmetics mandate that the products contain the materials specified in order to protect the consumer from product hazards as well as deceptive marketing practices. Quantitative analysis of vitamins in any system, whether they appear as intentional additives, pharmaceutical supplements, or dietary constituents, can and must be done on a regular basis by producer and marketer to monitor quality control. In the health care industry, tissue samples must also be measured in order to assess nutritional status. In the various stages of food-processing unit operations, when manipulations involve high temperatures and exposure to light, oxygen, and other nutrients such as metals, carbohydrates, and amines, labile vitamins may be destroyed. Further, storage conditions, rarely if ever optimal for retention of vitamin levels, typically result in losses of vitamins.

Analytical thin-layer chromatography (TLC) and high-performance TLC (HPTLC) are methods of choice for such evaluations and studies of vitamins because of their resolution, sensitivity, specificity, speed, and economy.

TLC methods have been developed to separate and measure all of the water- and fat-soluble vitamins. In contrast to current liquid chromatographic methods, several samples can be analyzed in one run on a single plate. Often 20 samples, including standards, can be chromatographed in one run. This often translates into reduced analysis time, on the average. Use of HPTLC with reduced solute loads also reduces analysis times to a few minutes rather than an hour in many cases. In addition, sample cleanup can be effected in part or whole in place. Most compounds in the chromatogram can be visualized after staining in the TLC system, which often reveals more information about the sample in terms of purity and degradation. Since most TLC plates are used only once and then disposed of, the cost of column replacement in case of fouling does not have to be considered. The sensitivity of most methods extends typically into the nanomole range, so that the methods can be applied to biological tissue samples, food, and pharamaceuticals.

The recent availability of automated sampling and spotting equipment, along with data reduction systems, makes rapid data generation available to most chromatography laboratories.

Reactions to modify the solute before detection can often be accomplished on the layer surface, and a wide variety of layers can be purchased, including standard silica gel G or H, silylated reversed phases, alumina, cellulose, polyamide layers, and combinations. The layers can be purchased in thicknesses ranging from 90 to 250 μm for analytical or preparative work. Glass, metals such as aluminum, and Mylar may be used as a backing support for the layer. The ready availability of a library of stains or other visualization techniques and of optical scanners has facilitated in situ quantitation of the vitamins and their derivatives.

Mobile phases to separate the vitamins and their derivatives have been available for both straight silica and reversed phases; several examples will be cited in this chapter to illustrate the potential uses of TLC in analytical chemistry of vitamins.

Detection of the vitamins has varied. Perhaps the most commonly applied method has been charring with an acid such as sulfuric or phosphoric, followed by reflectance or transmittance measurement (1). Direct fluorescence or quenched fluorescence has been used for several of the vitamins. Colorimetric methods (2, 3) have been used, as have radioimmunoassays of separated zones in the chromatogram, as in the case of vitamin D_3 (4). Recently, flame ionization detection has been used with chromatography on cylindrical rods coated with sorbent.

II. SELECTED METHODS FOR SEPARATION AND QUANTITATION OF VITAMINS BY TLC

A. Water-Soluble Vitamins

1. Thiamine

This vitamin has been measured in quality control, as in enrichment mixes, and in the clinical evaluation of beriberi, peripheral neural diseases, and Wernicke's encephalopathy commonly observed in alcoholism. The TLC is rapid and sensitive. Free thiamin, phosphorylated forms, and several analogs have been separated by TLC on polyamide layers (5) and silica gel (6). Separation and detection of 10-μg quantities of thiamin were accomplished with iodoplatinate (violet). Acetone or acetone-diethyl ether-glacial acetic acid (20:20:1, v/v), methyl ethyl ketone, and chloroform-ethyl acetate-glacial acetic acid (20:20:1, v/v) gave acceptable separations. The R_f values for nine forms of thiamin so separated are listed in Table 1. The mobile phases methyl ethyl ketone and chloroform-ethyl acetate-glacial acetic acid performed well.

2. Riboflavin

Thin-layer chromatography has been used to monitor riboflavin in vitamin production, food and vitamin sales, and clinical tests in dermatitis and neural disorders. Okumura et al. (2) reported the separation of several vitamins on sintered glass plates coated with silica gel G or H and a fluorescent substance. The authors noted that because of the inertness of the layer to corrosive acids (e.g., sulfuric, chromic, and trichloroacetic acids) and antimony trichloride, the versatility of these plates exceeds that of plates with organic binders. In an acetone-water solvent (9:1, v/v), riboflavin was separated from several other water-soluble vitamins having an R_f (× 100) of 47 on the sintered glass plate.

Separation and fluorometric quantitation of vitamin tablet preparations have been effected in citrate buffer (pH 2.0) on Kiesel gel HF 254 (E. Merck, Darmstadt, FRG) by Bican-Fister and Drazen (7). Riboflavin fluorescence was viewed under an ultraviolet (UV) lamp. To detect the material, scraped zones were ex-

Table 1 R_f Values for Thiamine and Derivatives[a]

	$R_f \times 100$		
Compound	I	II	III
Thiamine hydrochloride	5	5	2
Thiamine monophosphate	0	0	0
Thiamine disulfide	29	23	18
Thiamine propyldisulfide	72	62	74
Thiamine tetrahydrofurfuryl disulfide	68	59	67
Dibenzoyl thiamine	87	82	83
Thiamine β-hydroxyethyl disulfide	49	29	38
Dicethiamine hydrochloride	—	—	94
Thiamine 8-(methyl-6-acetyldi-hydrothioctate) disulfide	80	72	77

[a]Mobile phases: I, methyl ethyl ketone; II, chloroform-ethyl acetate-glacial acetic acid (20:20:1, v/v); III, acetone-diethyl ether-glacial acetic acid (20:20:1, v/v).
Source: Ref. 5.

tracted and fluorescence was recorded after dilution to 50 ml in water. Sensitivity was reported to be below 0.4 μg per spot when extracted, diluted, and scanned in this way.

Other methods have also been reported. In one novel medium, Petrovic et al. (8) separated riboflavin and other vitamins on thin layers of rice starch with *n*-propanol-pyridine-acetic acid-water. Kouimtzis and Papadoyannis (9) reported the separation of riboflavin in a mixture of five vitamins on layers of silica gel G. The mobile phase acetone-methanol-benzene (1:2:8, v/v) gave satisfactory separations. In situ fluorescence detection at 529 nm was used after excitation with a mercury lamp.

3. *Niacin: Nicotinic Acid/Nicotinamide*

Biotechnology research in which nucleic acids are manipulated in quantity and transferred between cells has necessitated the ability to assay nicotinic acid and nicotinamide rapidly. These applications to measure small amounts usually require highly sensitive methods. Food and pharmaceutical assays often require less sensitivity but the same resolution.

Frie et al. (10) demonstrated the separation of the acid and amide and their detection on cellulose layers. The eluant was butanol saturated with 3% ammonia. The chromatograms were developed in less than 1 h, after which they were selectively stained by the König reaction. This entailed mixing a saturated (ice-cold) bromine solution with an aqueous 10% sodium cyanide solution in a chamber until the color of the bromine disappeared. The chromatograms were then exposed to the saturated atmosphere of bromine cyanide for 1 h in the chromatography chamber. After that, the plates were sprayed uniformly with *p*-aminobenzoic acid (2 g) dissolved in hydrochloric acid (75 ml, 0.75 M), diluted to 100 ml with 96% (v/v) ethanol. Plates were dried for 1 h under a cold air stream. The zones appeared brown-yellow in visible light; fluorescence could be viewed at 254 nm. Fluorescence was quantitated with a 436-nm excitation filter and spectral maximum emission filters of 545 nm for nicotinamide at 530 nm for nicotinic acid. Reproducibility with an 8% relative standard deviation was obtained with six plates and was limited only by the uniformity of the spraying technique. The authors detected 1 ng of nicotinic acid and 2 ng of the amide per spot; sensitivity of the UV reflectance spectroscopy method was 0.1 μg per spot for either compound. Samples of cartilage red bone marrow extracts were separated and quantitated to demonstrate the utility of the method.

Wasshuttl (11) reported a special staining protocol for nicotinic acid and the amide after TLC. To develop the chromatograms, he used a mobile phase of *n*-propanol-aqueous ammonia (10%) (95.5, v/v) on layers of silica gel G (0.25 mm). The R_f values (× 100) were 35 and 65 for the acid and amide, respectively.

To stain the samples, the author sprayed the plate with 1-chloro-2,4-dinitrobenzene (2% in alcohol), after which the plate was heated at 180°C for 30 to 45 min. Under such conditions, the pyridine ring splits. Afterward, the plate is sprayed with alcoholic sodium hydroxide (3 N) at ambient temperature and then exposed to warm, concentrated ammonium hydroxide. As little as 0.2 μg of nicotinic acid appeared as a red-violet and nicotinamide as red-orange color. Either vitamin could be detected with linear calibrations attained from 0.25 to 2 μg. The method has a relative standard deviation of 10%.

Silica gel G has also been used to effect the resolution of nicotinamide in a citrate-buffered mobile phase, pH 2.0 (7). Sample eluted from spots was dissolved in 96% ethanol and absorbance was measured at 262 nm. A sensitivity of 250 μg per spot was reported in this spectrophotometric assay.

Gänshirt and Malzacher (12) reported separations from vitamin mixtures on silica gel G with a mobile phase of glacial acetic acid-methanol-benzene-acetone (5:20:70:5, v/v). As little as 1 μg could be detected. Bolliger and Stahl (13) reported similar separations with only water as the mobile phase.

4. *Pyridoxine*

The aldehyde pyridoxine and its phosphate were well separated on silica gel (14) after being first reductively aminated with methyl anthranilate in the presence of sodium cyanoborohydride to the anthranilates. The pyridoxines of serum were then quantitated fluorometrically.

Lyle and Tehrani (15) demonstrated the separation of vitamin mixtures containing pyridoxine hydrochloride on silica gel G-coated aluminum plates; a solvent phase of glacial acetic acid-acetone-methanol-benzene (5:5:20:20, v/v) was used. To see the spots, iodine vapors were reacted with the samples. The solute was quantitated by pyrolysis gas chromatography, measuring the acetaldehyde yield. The detector response to pyridoxines varied linearly with acetaldehyde.

In a similar separation on a silica gel G surface (fluor), Bican-Fister and Drazen (7) reported detection of the vitamin with UV light, under which pyridoxine appeared pale blue. Okumura et al. (2) used a mobile phase of acetone-water (9:1, v/v) to separate the pyridoxine in a seven-component mixture. They detected the vitamins by quenched UV fluorescence.

5. *Pantothenic Acid*

Silica gel G layers have been reported to separate the calcium salt of pantothenate from a mixture of pyridoxine hydrochloride and L-ascorbic acid (15). Glacial acetic acid-acetone-methanol-benzene (5:5:20:70, v/v) was used as the mobile phase to develop the chromatograms. To locate the spots, 0.1–0.5 M ammoniacal silver nitrate was sprayed on to wet the layers. Detection by pyrolysis gas chromatography was used to measure isobutanol, which was proportional to the vitamin itself. The authors reported that the sensitivity extended to 200 ng per spot.

6. *Folacin: The Folates*

Besides routine analysis of pharmaceutical preparations, clinical assessment of tissues for nutritional evaluation has necessitated assays of folic acid. Folate deficiency is widespread amond alcoholics and users of other licit drugs. Further, the efficacy of use of methotrexate, an antitumor analog with cytotoxic effects, has been improved with the availability of rapid, sensitive assays that enhance patient therapy with safety.

The separation of folates represents a particular challenge, since there are several methyl-substituted forms (methylene, methenyl, formyl, and formimino) and various oxidation states of the pteridine ring. Attachment of one to seven glutamate residues on the *p*-aminobenzoic acid further complicates chromatographic analyses. Conjugase has been used to cleave the glutamate residues prior to analysis, and this has also been done prior to microbiological assays (16).

Routine separation of folic acid and several derivatives by TLC has been reported by Reif et al. (3). On a fluorescent coated silica gel G, the authors developed samples dissolved in 0.3 N ammonium hydroxide from vitamin tablets with a mobile phase of 1.48 N ammonium hydroxide and *n*-propanol (1:3, v/v). Compounds which they separated by TLC, listed in order of increasing R_f values, included: folic acid; 2-amino-1,4-dihydro-4-oxo-6-pteridine carboxylic acid: *N*-(*p*-aminobenzoyl)-C-glutamic acid; and 2-amino-4)1*H*)-pteridinone. They eluted a sample from the gel, diluted in 0.17 M dipotassium phosphate, and detected it colorimetrically with a potassium permanganate-sodium nitrate-ammonium sulfamate-*N*-1-naphthylethylenediamine dihydrochloride solution (freshly prepared). Absorbance was recorded at 550 nm, yielding linear regression coefficients in the range of 50 to 200 ng of analyte. Quantitation was as good as that in HPLC, but the method was faster in spite of the elution step.

7. *p-Aminobenzoic Acid*

p-Aminobenzoic acid, a subunit of folacin, has also been separated (15). By using silica gel G, this vitamin component was separated from a mixture of thiamine, hydrochloride, riboflavin, pyridoxine hydrochloride, and nicotinamide with a mobile phase of glacial acetic acid-acetone-methanol-benzene (5:5:20:20, v/v). The diazo method, with the Marshall reagent mentioned above containing *N*-naphthylethylenediamine dihydrochloride, gave satisfactory quantitative results at 545 nm.

8. *Cobalamin*

Like the folic acid group, the corrinoids present a rigorous analytical challenge. Pharmaceutical and medical interest in them has

Table 2 R_f Values of Corrinoids Separated on Cellulose Layers

Axial ligand-Cobalamin	$R_f \times 100$
1. 2,2-Diethoxyethyl	62
2. Methyl	69
3. 1,3-Dioxa-2-cyclopentylmethyl	56
4. Formylmethyl	51
5. Cyano	48
6. 2,3-Dihydroxypropyl	46
	38
7. 5-Deoxyadenosyl	34
8. Hydroxo	22

Source: Ref. 18.

focused on their use in treating megaloblastic anemia, which is widespread today.

Tortolani and Ferry (17) separated cobalamins on layers of silica gel G. Silverman and Dolphin (18) separated several axial substituted cobalamins on cellulose layers. The mobile phase *n*-butanol-ethanol-water (containing 0.5% concentrated aqueous ammonia (10:3:7, v/v) effected satisfactory separations of samples dissolved in 0.5% concentrated aqueous ammonia at ambient temperature. The R_f values for eight axial ligand substituted cobalamins are shown in Table 2.

9. *Inositol*

Pharmaceutical preparations have been routinely analyzed for the cyclic alcohols such as inositol by using high-performance thin layers of silica gel G (19). Acetone-methanol-water (2:2:1, v/v) effected separation from other sugars; fluorescence with lead tetraacetate/fluorescein allowed detection of 0.5 ng per spot.

10. *Biotin*

In quality control and lipid metabolic disease tests, analysis for biotin has been done by TLC, radioimmunoassay, and microbiological assay. Gänshirt and Malzacher (12) separated biotin in vitamin mixtures with a mobile phase of acetic acid-acetone-methanol-benzene (5:5:20:70, v/v) on Kiesel gel G. In this assay they could detect 1 μg per spot with an iodoplatinate stain.

11. Ascorbic Acid

Widespread interest in ascorbic acid has been based on its sales as a health food and pharmaceutical in recent years; it is perhaps the most popular vitamin among consumers in the United States. Specialty applications of this reducer have included food uses as an antioxidant in parenteral feeding solutions and uses in flavor systems and compounded vitamin and mineral tablets. It has appeared in other specialty chemicals as well, including baked products. Measurement of ascorbate has been based largely on titration in liquid systems, oxidation and potentiometric detection, or formation of the bis-osazone and colorimetric determination.

Hoppe et al. (20) separated the osazone derivatives by TLC from whole milk that had been deproteinated. The authors coupled 2,4-dinitrophenylhydrazine to the ascorbate molecule and measured the substrate in situ with a densitometer. The calibration curve absorption values were linear from 20 to 250 μg/ml in milk. The coefficient of variation was 2.3 ± 1.5% in replicate samples. Recovery was 95%.

Matthiessen (21) used a similar approach to separate and quantitate ascorbic acid and dehydroascorbic acid after its oxidation to ascorbic acid. The author did a preliminary cleanup on a silica gel column and separated the derivative on silica gel G. The red-orange hydrazone spot was detected in situ with a densitometer. Samples and calibration curves run on each plate (from standard concentrations of ascorbic acid) were scanned at 500 nm. The method's sensitivity extended to 0.2 mg/L, which was demonstrated with UHT pasteurized milk samples.

B. Fat-Soluble Vitamins

1. Retinol/Retinoic Acid

Quantitation of retinol and its analogs has been important in connection with two main diseases: widespread blindness in starving children of developing countries, and a recently noted inverse relation between tissue retinoids and cancer risk (22, 23). Quality control and other medical uses also exist.

To monitor the quantity and quality of retinoids in topical creams and gels, De Paolis (24) devised a two-dimension high-performance TLC system. In particular, a rapid system is needed to analyze all-*trans*-retinoic acid, since brief exposure to light (550 nm) isomerizes the vitamin. To analyze gels having 0.01% solute, a methanol extract was applied directly to a plate, and the plate was developed with diethyl ether-cyclohexane-acetone-glacial acetic acid (40:60:2:1, v/v). The chromatography, carried out in foil-wrapped tanks, could be done in 10 to 15 min. Sensitivity was 20 ng for either isomer. The sytem satisfactorily separated the all-*trans* form from the 13-*cis* form (stored in methanol with 0.5% butylated hydroxytoluene to retard oxidation). Cream preparations having 0.05–0.1% solute contained interfering excipients that necessitated cleanup by reversed phase chromatography. By use of a two-phase plate having both normal and reversed phase surfaces, the sample could be developed in ethanol-distilled water (80:20, v/v) on the reversed surface, dried, rotated 90°, and developed normally with diethyl ether-cyclohexane-acetone-glacial acetic acid (40:60:2:1, v/v). All manipulations were done under subdued light; samples were sprayed with a solution of concentrated sulfuric acid in ethanol (8:92, v/v). A detection limit of 20 ng was reported for each isomer. By spotting reference material to the side of the plate, the degree of isomerization during analysis could be monitored.

Kahan (25) reported chromatography of retinol metabolites extracted from human serum and liver samples on silica gel G layers. The free retinol, retinyl acetate, and retinyl palmitate were isolated from the serum and liver samples. Three solvent systems resolved the compounds with the following respective R_f values: petroleum ether-ethyl ether-acetic acid (90:10:1, v/v), 0.08, 0.65, 0.85; cyclohexane-ethanol (97:3, v/v), 0.08, 0.37, 0.85; and chloroform-ethanol (99:1, v/v), 0.24, 0.77, 0.86. Compounds were detected by fluorometry, UV absorption, or the antimony trichloride color reagent (20% $SbCl_3$ in chloroform). Samples were eluted and absorption and fluorescence spectra recorded.

2. *Provitamin A: Carotenoids*

Rapid separation of natural carotenoid pigments was reported by Keefer and Johnson (26). After trials with silica gel G, they turned to magnesium hydroxide thin layers; for masses over 20 mg, they used columns. They reported satisfactory separations for the α and β isomers of carotene, and for lycopene, in carbon disulfide. The R_f values for α-carotene, β-carotene, and lycopene were 0.43, 0.29, and 0.02, respectively. Solvents such as pyridine, tetrahydrofuran, and ethanol did not promote effective separation on magnesium hydroxide, probably because this solvent

Table 3 R_f Values of Carotenoids and Retinoids in Indicated Solvents on Magnesium Hydroxide Layers

	$R_f \times 100$		
Compound	Benzene	Carbon tetrachloride	Carbon disulfide
α-Carotene	—	—	43
β-Carotene	—	—	29
Lycopene	—	—	2
Retinol	29	5	—
Retinal	61	22	—
Retinyl acetate	75	48	—

Source: Ref. 26.

promoted an interaction with a polar group, increasing the absorption energy. Polar substituents such as a hydroxyl group or an ester, as in retinol and retinyl acetate or retinaldehyde, also increased the absorption energy. See the R_f values in Table 3. The retarding effect of carbon disulfide was also apparent in these data. The number and location of double bonds also influenced the migration of these compounds.

Separation of several carotenoids was reported by Cecchi and Rodriguez-Amaya (27) who used a mixture of magnesium oxide and Hyflosupercel (1:2, v/v) in columns with a gradient of ether and acetone in petroleum ether. They separated the same compounds on silica gel G layers with 3% methanol in benzene. The carotenes (C) separated included α-C, β-C, ζ-C, *cis*-β-C, cryptoxanthin, aurochrome, cryptochrome, auroxanthin, phytoene, and phytofluene. Absorbance data were used to identify and quantitate the compounds, except for the last two named, which are devoid of vitamin activity in vivo.

3. *Calciferols: Provitamins and Vitamins D*

Provitamins and vitamins D have become important in prevention or therapy of bone disease. Analytical TLC has been chosen for its speed and sensitivity, since these compounds decompose rapidly in air or light. Silica gel G and alumina GF effectively separate many sterols, besides the vitamins. Ponchon and Fellers (28) used mobile phases of chloroform or petroleum ether-benzene (v/v) to separate the sterols, followed by any of several detection methods. The authors separated cholesterol, β-sitosterol, stigmasterol, lanosterol, ergocalciferol, pre-ergocalciferol, cholesteryl acetate, cholesteryl benzoate, and stigmasterol acetate with either of these mobile phases. On silica gel G layers, *n*-hexane-acetone (95:5, v/v) also yielded satisfactory separations.

Several detection methods have been used for detection and quantitation. The sterols were detected well by fluorescence quenching at 225 nm or excitation at 366 nm: these included ergosterol, ergocalciferol, cholecalciferol, previtamin D, lanosterol, and cholesterol ester. Charring with 75% sulfuric acid heated at 80°C yielded green-violet colors, but if held at 150°C the fluorescence (excited with longwave UV) increased so that detection limits were below 0.5 μg per sample. Antimony trichloride followed by heating at 110°C for 20 min converted the analytes into highly fluorescent spots. Typical sensitivity was well below 1 μg per spot. Colorimetric reactions, such as reaction with phosphotungstic acid or phosphomolybdic acid, gave acceptable linear regressoins. Sensitivity was 0.5 to 1 μg per spot at best.

Of more interest recently, the hormone (hydroxylated) derivatives of the vitamins D have presented another set of separation problems. The ability to separate and quantitate them has become very important in clinical and pharmaceutical laboratories, since certain of these derivatives have been prescribed for patients suffering with bone disease, such as osteomalacia, or renal osteodystrophy.

In addition to testing the purity of preparations, dosage monitoring has become very important since overdosage introduces another lesion. Justova et al. (4) reported separating the hydroxy metabolites on silica gel G with chloroform-ethanol-water mobile phase (183:163:1, v/v). Vieth et al. (29) used a hexane-isopropanol system (85:15, v/v).

A newer high-performance TLC (HPTLC) method has been published (30). In contrast to conventional TLC, this HPTLC system required much less sample material, nanograms rather than micrograms, and less time, 11–12 min rather than 35–40 min. The high-performance silica gel plates (10 × 10 cm) Analtech, Newark, DE) required one-tenth as much solvent, 20 ml, as a conventional tank for 20 × 20 cm plates. Comparative retention data for the compounds in three solvent systems appear in Table 4. From such retention data, one would choose a mobile phase separating compounds from each other, or the origin, as desired.

Table 4 Chromatographic Data for Several Vitamin D Metabolites on Silica Gel by HPTLC

	$R_f \times 100$		
Compound	Dichloromethane-isopropanol (90:10)	Chloroform-ethyl acetate (50:50)	Hexane-isopropanol (85:15)
$1,24,25(OH)_3D_3$	21	05	14
$1,25(OH)_2D_2$	37	19	26
$25,26(OH)_2D_3$	44	29	35
$24,25(OH)_2D_3$	61	44	46
$23,25(OH)_2D_3$	64	51	51
$25(OH)D_3$	73	71	59
Vitamin D_3	89	87	66

Source: Ref. 30.

Food labeling laws require honest declarations of vitamin content, another application of TLC of vitamins. In one application to evaluate the level of vitamin D in dietetic foods, Meine (31) used TLC on silica gel to separate and quantitate the vitamin. After extraction of fat and fat-soluble vitamins with ligroine and ethyl ether, the extract was saponified and nonsaponifiables were extracted. Meine (31) chromatographed these on Florex-XXS, then washed the column with *n*-hexane; vitamin D was eluted with toluene. After chromatography with cyclohexane-ethyl acetate (3:1, v/v), the vitamin D was quantitated by reflectance spectrometry at 265 μm. The TLC values for instant milk products (428 IU/130 g) agreed satisfactorily with results obtained with a biological method (465 IU/130 g).

4. *Tocopherols (Vitamin E)*

Important in numerous food system applications as natural antioxidants (32), in clinical disorders such as cystic fibrosis (33) and epilepsy (34), and in numerous cosmetics and other pharmaceuticals, the several tocopherols must be separated since different forms have distinct biological and general antioxidant potentials.

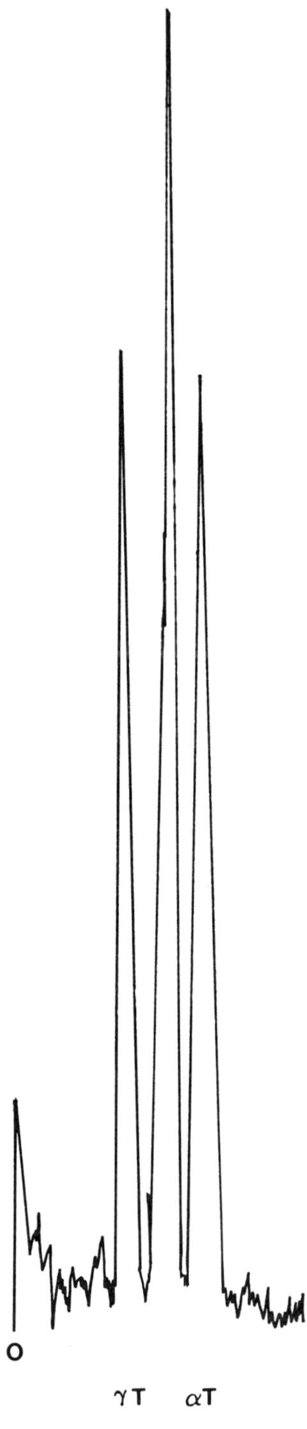
O
γT
αT
αTT

Whittle and Pennock (35) reported a two-dimensional mobile phase system for separating several of the tocopherols (T), including α-T, α-tocotrienol, β-T, β-tocotrienol, γ-T, γ-tocotrienol, δ-T, and δ-tocotrienol. Detection, based on the Emmerie-Engel reaction of ferric chloride in dipyridyl solution, required elution of the sample and absorbance readings at 520 nm. An alternative method entailed use of an ethanolic fluorescent spray and visualization with UV light (270–280 μm), under which spots appeared purple on a fluorescent background. As little as 1–2 μg per spot could be detected by this method on silica gel G.

Aratani et al. (36) separated tocopherols on very thin (70-μm) layers of silica gel G bound to etched quartz microscope slides. Several surfaces and mobile phases were used: benzene or benzene-ethyl ether (v/v) on alumina; benzene or chloroform on silica gel G; chloroform on alumina-zinc carbonate (3:1, v/v); benzene-cyclohexane (30:70, v/v) on silica gel G-zinc carbonate (2:1, v/v); and chloroform followed by isopropyl ether-light petroleum (20:80, v/v) on silica gel G. The authors demonstrated that separations of submicrogram quantities could be attained in 13 min by using layers 53 nm thick on silica gel G with a solvent of *n*-hexane-chloroform (50:50, v/v). In situ quantitation was done with a spectrophotometer in the transmittance mode (Shimadzu, Osaka).

In studies of algal synthesis of tocopherols, Ruggeri et al. (37) used TLC on silica gel G to separate several methyl forms of tocopherol (α-T, β-T, α-tocotrienol, γ-T, and δ-T) and the unsubstituted tocol form. One satisfactory mobile phase was *n*-hexane-isopropyl ether on silica gel G (Analtech, Newark, DE); methanol-water (95:5, v/v) on (C18) reversed phase surfaces (EM Science, Morristown, NJ). Figure 1 shows a representative chromatogram. This system had linear response with in situ quantitation of spots charred with 3% cupric sulfate in 8% phosphoric acid, with viewing at 350 nm by reflectance. Sensitivity was 200 ng per spot. It was chosen over a liquid chromatography system because of its speed, solvent economy, and simplicity. Examples of food tocopherol chromatograms obtained by this method appear in Figures 2 and 3, which show distinct synthesis of α-T by papaya and γ-T by mango. Tocopheryl esters such as the acetate and palmitate can be separated easily by this system.

Figure 1 Chromatogram of tocopherol reference compounds: α-tocopherol (α-T); γ-tocopherol (γ-T); α-tocotrienol (α-TT). See text for conditions.

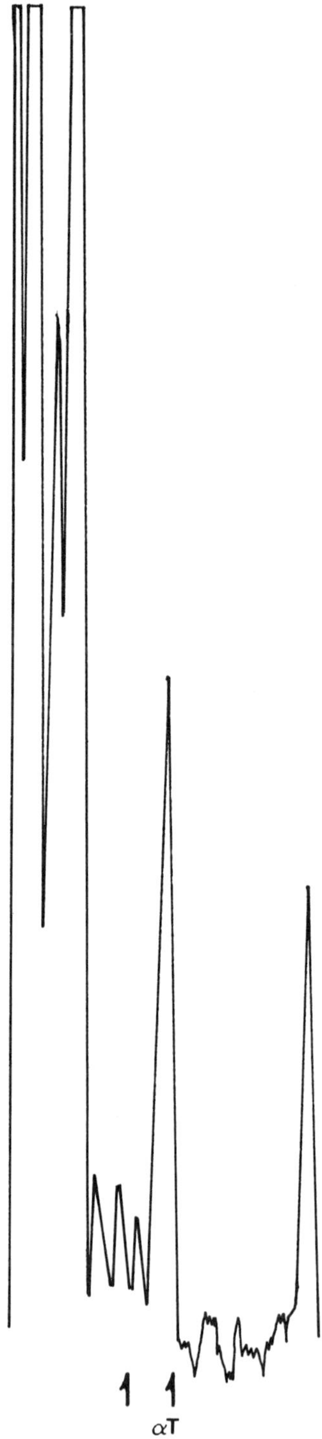
αT

5. *Quinones (Vitamin K)*

Menaquinones (K2–K9), phylloquinone (K1), and menadione have been separated readily by TLC. Clinical uses of these materials include postsurgical hyperalimentation solutions, pharmaceutical capsules, foods, and plant extracts. Because of the essentiality of the vitamins K in γ-glutamyl carboxylation reactions and blood clot formation, they comprise a group of cofactors that must be monitored routinely in physiologically critical situations.

Thin-layer chromatography has been used as a practical technique to separate the isoprenologues as well as the major types of naphthoquinones. Hammond and White (38) separated *Streptococcus aureus* isoprenologues having 0 to 50 carbon atoms in the side chain with the mobile phase of acetone-water (95:5, v/v) saturated with hexadecane. The thin layers, specially prepared, consisted of Kieselguhr G, 50 μm, impregnated with hexadecane (5% in chloroform, ascending). To detect the samples, the plates were sprayed with 1% sodium borohydride in 50% ethanol, followed by 2% aqueous neotetrazolium (dissolved in 4 ml of 95% ethanol and made up to 200 ml in water). The quinones appeared as red spots without heating. The authors quantitated by counting incorporated radiolabeled mevalonate in the cell extracts.

Dunphy et al. (39), in comparative studies of the microbial synthesis of quinones, separated 13 menaquinones from several organisms (*M. phlei, C. createnovorans, M. tuberculosis, Streptomyces* spp., *M. lysodeikticus, Flavobacterium* spp., and *B. stearothermophilus*). When silica gel G layers were used, solvents were *n*-butyl ether-hexane (10:90, v/v) or ethyl ether-pentane mixtures; for silver-impregnated silica gel G layers, solvents were methanol-benzene (5:95, v/v) or 2-butanone-hexane or acetone-pentane (10–20%, v/v); and for reversed phase layers, solvents were dimethyl formamide-water (97:3, v/v) or acetone-water (90–97%). The reversed phase systems were saturated before use with liquid paraffin. The menaquinones were identified and quantitated by UV absorption maxima.

Thielemann (40) reported that 4-aminophenazone condenses readily with the major naphthoquinones (K1, K2, and K3) to form a bright red adduct. These compounds can be chromatographed on silica gel G plates with a mobile phase of benzene-acetone (90:10, v/v).

Figure 2 Chromatogram of extract of papaya seed lipids. Note preponderance of α-tocopherol. See text for conditions.

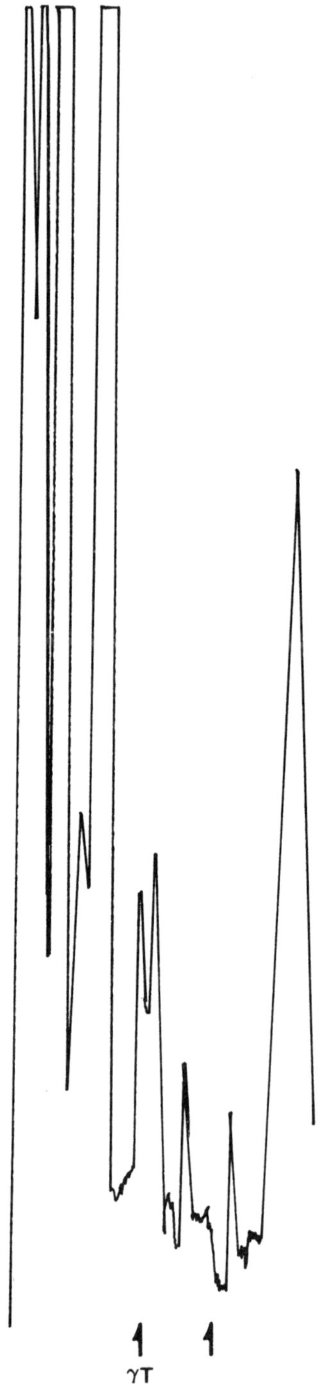
γT

In a study of the decomposition of vitamins exposed to ultraviolet light, Baczyk et al. (41) studied vitamins D2 and K1. They separated these vitamins on silica gel H with chloroform as the mobile phase. After visualization of the chromatograms with fluorescein, samples were eluted with ethanol and quantitated by absorption spectra at 265 and 248 nm, respectively.

III. SUMMARY

Normal and reversed phase techniques have been and continue to be used to quantitate vitamins, in both industrial and clinical laboratories. Numerous examples have been presented. Although formerly quantitation often entailed eluting the analyte from the gel followed by spectrophotometry, currently developed methods generally utilize in situ quantitation, whether colorimetric, fluorometric, or radiometric. In many protocols, sample cleanup has been accomplished on the surface in a preadsorbent or reversed layer. The sensitivity of TLC methods, especially HPTLC, compares favorably with that of high-pressure liquid chromatography. Analysis with TLC may be considerably faster, since multiple samples can be analyzed simultaneously. Derivatization or detection reactions in TLC have typically been done on the plate without accessory equipment. The time savings correspond to money saved. The speed, sensitivity, reproducibility, and automation achievable with off-the-shelf TLC plates and densitometers suggest that quantitative TLC will continue to be a versatile and widely used analytical technique for measuring the vitamins.

Figure 3 Chromatogram of extract of mango seed lipids. Note that γ-tocopherol is the major form of tocopherols. Conditions are described in the text.

REFERENCES

1. T. R. Watkins, B. Ruggeri, R. J. H. Gray, and R. I. Tomlins. In: *Techniques and Applications of Thin Layer Chromatography*, J. C. Touchstone and J. Sherma (eds.), Wiley, New York, 1985.
2. T. Okumura, T. Kadono, and M. Nakatani. J. Chromatogr. *74*, 73 (1972).
3. V. D. Reif, J. T. Reamer, and L. T. Grady. J. Pharm. Sci. *66*, 1112 (1977).
4. V. Justova, Z. Wildtova, and L. Starka. J. Chromatogr. *230*, 319 (1982).
5. H. P. Chuang, H. C. Chiang, and K. T. Wang. J. Chromatogr. *41*, 487 (1969).
6. C. Levorato and L. Cima. J. Chromatogr. *32*, 771 (1968).
7. T. Bican-Fister and V. Drazen. J. Chromatogr. *77*, 389 (1973).
8. S. E. Petrovic, B. E. Belin, and D. B. Vakujlovic. Anal. Chem. *40*, 1007 (1968).
9. Th. A. Kouimtzis and I. N. Papadoyannis. Mikrochim. Acta *2*, 145 (1979).
10. R. W. Frie, A. Kunz, G. Pataki, T. Prims, and H. Furcher. Anal. Clin. Acta *49*, 527 (1970).
11. J. Wasshuttle. Mikrochim. Acta (1970).
12. H. Gänshirt and A. Malzacher. Naturwissenshaften *47*, 279 (1960).
13. H. R. Bolliger. In: *Thin Layer Chromatography*, E. Stahl (ed.), p. 237, Academic Press, New York, 1965.
14. M. S. Chauhan and K. Dakshinamurti. Anal. Biochem. *96*, 426 (1979).
15. S. J. Lyle and M. S. Tehrani. J. Chromatogr. *236*, 31 (1982).
16. E. L. R. Stokstad and J. Koch. Phys. Rev. *47*, 83 (1967).
17. G. Tortolani and P. G. Ferry. J. Chromatogr. *88*, 430 (1974).

18. R. B. Silverman and D. Dolphin. J. Chromatogr. *194*, 273 (1980).

19. J. Stepanek. J. Chromatogr. *257*, 405 (1983).

20. A. Hoppe, E. H. Reimerdes, and H. Reuter. Milchwissenschaft *34*, 263 (1979).

21. E. Matthiessen. Dtsch. Lebensm. Rundsch. *74*, 403 (1978).

22. M. B. Sporn. Nutr. Rev. *35*, 651 (1978).

23. G. Wolf. Nutr. Rev. *40*, 257 (1982).

24. A. M. De Paolis. J. Chromatogr. *258*, 314 (1983).

25. J. Kahan. J. Chromatogr. *30*, 506 (1967).

26. L. K. Keefer and D. E. Johnson. J. Chromatogr. *69*, 215 (1972).

27. H. M. Cecchi and D. B. Rodriguez-Amaya. J. Food Sci. *46*, 147 (1981).

28. J. Ponchon and F. X. Fellers, J. Chromatogr. *35*, 53 (1968).

29. R. Vieth, D. Fraser, and G. Jones. Anal. Chem. *50*, 2150 (1978).

30. M. Thierry-Palmer and T. K. Gray. J. Chromatogr. *262*, 460 (1983).

31. W. Meine. Z. Lebensm. Unters. Forsch. *170*, 251 (1980).

32. J. C. Bauernfeind. Int. J. Vitam. Nutr. Res. Suppl. *27*, 307 (1985).

33. P. Ferrell, J. Bieri, J. Frantantoni, R. Wood, and P. diSant 'Agnese. J. Clin. Invest. *60*, 237 (1978).

34. A. O. Ogunmekan. Am. J. Clin. Nutr. *32*, 2269 (1979).

35. K. J. Whittle and J. F. Pennock. Analyst *92*, 423 (1967).

36. T. Aratani, K. Mita, and F. Mizui. J. Chromatogr. *79*, 179 (1973).

37. B. A. Ruggeri, T. R. Watkins, R. J. H. Gray, and R. I. Tomlins. J. Chromatogr. *291*, 377 (1984).

38. R. K. Hammond and D. C. White. J. Chromatogr. *45*, 446 (1969).

39. P. J. Dunphy, P. G. Phillips, and A. F. Brodie. J. Lipid Res. *12*, 442 (1971).

40. H. Thielemann. Mikrochim. Acta 227 (1972).

41. S. Baczyk, L. Duczmal, I. Sobisz, and K. Swidzinska. Mikrochim. Acta *2*, 151 (1981).

8
Analysis of Alkaloids

PÉTER HORVÁTH Chemical Works of Gedeon Richter, Ltd., Budapest, Hungary

I. INTRODUCTION

The naturally occurring nitrogen-containing compounds with strong physiological effects, isolated from different plants, were given the name *alkaloids* by Meissner in 1818 for their basic chemical character. The alkaloids are grouped according to numerous criteria, and the borders of the classes of alkaloids are drawn differently from case to case. In terms of their biosynthesis in plants the alkaloids are derived from some amino acids or closely related compounds, and they can be classified as alkaloids derived from (a) ornithine and lysine, (b) tyrosine and phenylalanine, (c) tryptophan, (d) anthranilic acid, and (e) histidine. Mainly for traditional reasons (f) steroid alkaloids, (g) xanthine alkaloids, and (h) miscellaneous groups are discussed here, with the quaternary ammonium salts among them.

The alkaloids and their chemical analysis are most important in the pharmaceutical industry. However, the quantities of alkaloids are investigated more and more frequently in different food industrial products and in materials of the tobacco industry.

The pharmaceutical industry uses more than 100 alkaloids in different preparations. In this industry the usual aims of quantitative analysis of alkaloids are as follows:

1. Determination of the quantity of alkaloids occurring in fresh or dried plant materials (fermented fluids constitute an exception) for the purpose of plant breeding and/or processing
2. Investigation of the intermediates and starting materials for the purpose of process control and development
3. Quality control of pharmaceutical products.
4. Optimization of reaction parameters in the preparation of new semisynthetic or synthetic compounds
5. Quality control of new experimental materials prior to their pharmacological investigation.

II. GENERAL QUANTITATIVE TLC METHODS FOR ALKALOIDS

In the chemical analysis of alkaloids quantitative thin-layer chromatography (QTLC) plays an important role and is an alternative to high-performance liquid chromatography (HPLC). Gas chromatography is used primarily in the field of tobacco alkaloids.

The alkaloids, disregarding a few exceptions, are heterocyclic compounds containing tertiary or quaternary nitrogen atoms. Their basicity is usually weak in case of an *N*-heteroaromatic ring ($pK_b = 7 - 14$). However, when the nitrogen atom is part of a saturated ring, the basicity is much stronger, similar to that of the appropriate aliphatic amines. The tertiary and quaternary alkaloids in acidic media can be precipitated with a number of heavy metal ions as complexes, and some of these can be used as visualization reagents for the detection of the compounds, which are generally colorless and frequently do not absorb in the ultraviolet (UV).

The visualization reagents commonly used for alkaloids are different varieties of Dragendorff's reagent with $H(BiI_4)$ and the hexaiodoplatinate reagent with $H_2(PtI_6)$ as active ingredients. Only the alkaloids containing tertiary and quaternary nitrogen give positive reactions with these complexes. However, with in situ postchromatographic alkylation the detection of primary and secondary amines can also be facilitated by using these reagents (1).

The xanthine alkaloids cannot be detected directly with these complexes, except for those with an amino group in the side chain. However, Hauck (2) found that these compounds can also be detected when the plate is first sprayed with silver nitrate in 5% sulfuric acid before Dragendorff's reagent is used. A similar effect is obtained by previous spraying with 20% nitric acid and heating (3).

Different modifications of Dragendorff's reaction (4–9) for in situ QTLC purposes are customary. Application of this reaction is especially advantageous where a similarly sensitive alternative detection method is not known. With Dragendorff's reagent modified according to Munier (4), the alkaloids give insoluble precipitates in the reagent solution and the reaction is fast enough not to cause any observable spreading of spots during the detection.

Figure 1 shows densitograms of *Catharanthus* alkaloids obtained (a) without and (b) with visualization by Dragendorff's reaction modified according to Munier (4). The separation of the neighboring components by the postchromatographic reaction is not worse than by direct reflectance spectrophotometry. A similar result was obtained with the hexaiodoplatinate reaction.

For the detection of alkaloids on TLC plates, different varieties of Dragendorff's reagent are used. These modifications differ from each other in the concentration of the iodobismutate complex, the quality of the solvents, the quality of the acid used, and the treatment subsequently applied. In QTLC practice, Munier's modification (4) usually seems to be applied, but in special cases other varieities may also have advantages. Posttreatment with sodium nitrate (9) decolors the background; it decomposes the $(BiI_4)^-$ complex while I_2 evolves. However, the alkaloid spots also become paler, and in a more or less short time they usually disappear. Care must also be taken because some solvents, e.g., dimethyl formamide and formamide, leave residues on the plate that interfere with Dragendorff visualization (10). In the case of formamide-impregnated plates, formamide can be decomposed in situ with nitrous acid.

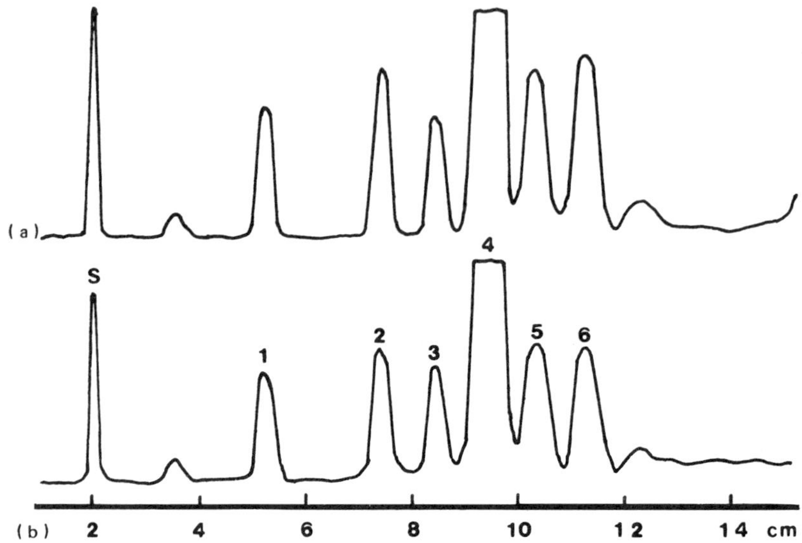

Figure 1 Densitograms of *Catharanthus* alkaloids. (a) Direct densitometry at 254 nm. (b) Densitometry at 400 nm after Dragendorff's reaction. Scanning, Opton KM 3 chromatograph-spectrophotometer, reflectance mode; layer, silica gel 60; mobile phase, ethyl acetate-benzene-methanol-25% ammonia solution (100:5:5:5, v/v). 1, Vincristine; 2, anhydrovinblastine; 3, *N*-desmethylvinblastine; 4, vinblastine; 5, leurosine; 6, desacetoxyvinblastine.

For QTLC purposes in general, and for alkaloids as well, the dipping technique gives better quantitative results than spraying with the reagent (11).

The tetraiodobismutate reaction gives different sensitivities in the detection of different groups of alkaloids. This reaction can be used especially advantageously for the investigation of certain pharmaceutical products and related compounds belonging to the quaternary ammonium salts, which have no chromophore group to make detection easier. In this group Dragendorff's reagent permits the detection of as little as 0.1% of contaminants besides the main component. In this case the calibration curve is linear without any transformation in the region from 0.05 to 1.0 μg. (See conditions in Figure 4.)

Another general visualization reagent for the alkaloids is iodine in vapor form or in solution. It can be used for QTLC purposes as well, if the plate with the visualized chromatogram is hermetically covered with a glass plate to prevent evaporation of the iodine adsorbed on the spots. In the case of QTLC of quaternary ammonium compounds mentioned above, the iodine reaction gives significantly less reliable results than the iodobismutate reaction.

In conventional TLC there is no effective universal detection method similar to flame ionization detection (FID) used in gas chromatography. In quality control of different industrial preparations it is frequently important to know that the material examined does not contain any organic impurity above a certain concentration limit. The microrod TLC technique (12), which has been successfully applied with alkaloids, can combine the special advantage of TLC, namely that all components chromatographed are necessarily on the chromatogram, with the unique feature of FID of being able to detect practically every organic compound. Unfortunately it has not been possible to use FID with regular chromatoplates. The microrod technique, on the other hand, has lost some of the advantages of conventional TLC.

Many attempts have been made to find a general detection method for conventional TLC plates. Segura and Gotto (13) used drastic heating in the presence of ammonium hydrogen carbonate and Bochoff et al. (14) used ionization to convert a great number of different compounds, among them alkaloids, on the plate to measurable fluorescent derivatives. In other cases, exposure to the vapor of red-fuming nitric acid followed by high-temperature treatment gave a good detection method for compounds detectable with difficulty (15).

In certain instances severe reaction conditions are used for the detection of the alkaloids that do not contain a chromophore or fluorophore group. For example, Karlsson and Peter (16) worked out a fluorodensitometric method for the detection of *Lupinus* alkaloids and used heating at 130°C for as long as 17 h for the detection.

A great number of alkaloids naturally absorb in the UV or visible region or fluoresce, and these compounds can easily be detected directly. Some alkaloids, partly because of their easy detection, are even popular as model compounds for demonstrating the effectiveness of new chromatographic techniques.

In many cases alkaloids are not directly detectable and can be determined better with special reagents than with general ones. In the investigation of different groups the alkaloid-specific reagents are well known, and many of them are suitable for quantitative TLC purposes as well.

In TLC the postchromatographic reactions can usually be accomplished more easily than in HPLC, so that prechromatographic reactions do not have any great importance in TLC. However, in many instances the transfer of some well-established precolumn reactions into TLC methods is advantageous. In the field of alkaloids Frei et al. (17), among others, used this obvious possibility.

One of the prerequisites of reliable quantitative TLC is suitable separation of the components to be determined. In general, the selection of appropriate solvent systems is somewhat more tedious for alkaloid determinations than for determinations of other organic compounds, because of the higher polarity and basic character of alkaloids. However, in recent decades many effective TLC systems have been designed for the separation of different groups of alkaloids and are described in the appropriate chapters of books on TLC (18–24).

The other factor in the efficiency of separation is the chromatographic technique applied. Recent developments in TLC techniques and their separation efficeincy have affected the field of alkaloid investigations. In a number of cases high-performance TLC (HPTLC) plates are used, which give separations that are better or are similar to those with conventional paltes in a shorter migration distance and with more reliable quantitative data. Among the other recent technical developments, overpressurized TLC (OPTLC) (25–27) seems to excel; this is the first forced-flow technique in TLC, and it has the ability to control and optimize the solvent velocity during chromatography. Although OPTLC has its unique advantages, it does not automatically lend itself to transfer of the conventional chromatographic systems. A significant improvement in separation with OPTLC occurs only when the other chromatographic conditions, e.g., the particle size of the stationary phase, are not suited for conventional developing conditions.

Nevertheless, in some areas of the alkaloids OPTLC has already been used with success. For example, Gulyás and Mincsovics (28) separated several poppy alkaloids by the OPTLC technique. The separation shown in Figure 2 was carried out in a Chrompres 10 chamber (Labor MIM, Budapest).

The high separation efficiency can be important, especially for naturally occurring biological materials, such as natural alkaloids, where great care is necessary to avoid errors caused by overlapping of the spots to be measured with those of other compounds. A simple method was applied by Karlsson and Peter (16) in a QTLC procedure for screening of alkaloids of *Lupinus* species in order to develop a breeding program. To control the efficiency of the chromatographic system, a two-dimensional chromatogram

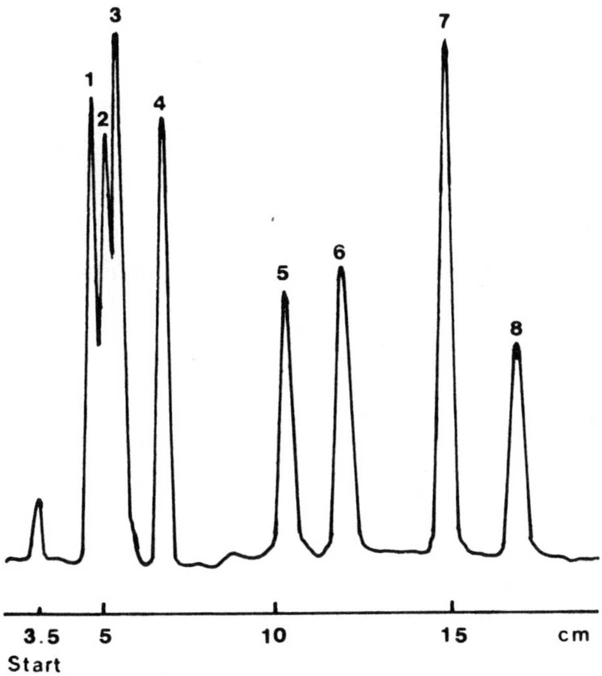

Figure 2 Separation of poppy alkaloids by OPTLC. 1, Morphine; 2, normorphine; 3, narceine; 4, codeine; 5, narcotoline; 6, thebaine; 7, papaverine; 8, narcotine; S, start. Layer, HPTLC silica gel 60 F-254; mobile phase, chloroform-acetonitrile-propionic acid-water (15:5:7:0.5, v/v); cushion pressure, 1.0 MPa; flow velocity, 0.18 ml/min; time of development, 20 min (6 min over-running); densitometry, Shimadzu CS 920 Scanner, absorbance measurement at 230 nm.

was identical to that in the one-dimensional method. Although this does not constitute conclusive proof, the likelihood of a clear separation for the main components is significantly increased. The control method is frequently HPLC; e.g., Horváth et al. (29) used an elution QTLC method for the determination of ergot alkaloids, and an HPLC procedure for the control assay.

A great many of the alkaloids are optically active compounds, and it is a special problem to investigate the optical isomers in their mixtures. The separation of members of diastereomer pairs does not differ, in principle, from that of any other difficult-to-separate compound pairs. However, the members of enantiomer pairs can be separated only with optically active chromatographic systems. Unfortunately, for most of the separation problems such systems with suitable efficiency have not been available. Another possible method of separation involves prechromatographic reaction with an optically active reagent that converts the enantiomers into separable diastereomer pairs. Eskes (30) separated the *d*- and *l*-cocaines after their conversion to the diastereomer ecgonine-2-octanols.

Horváth et al. (31) used another method for the determination of the four optical isomer lysergic acids in their mixtures. They did not attempt a total separation, and they used a simple, and not optically active, chromatographic system in which the members of the enantiomer pairs do not separate from each other. They supplied the missing analytical information from the data from a multiwavelength optical rotatory dispersion (ORD) measurement. In this case the chromatographic task simplifies the separation problem to half as many components. The simplification achieved with a combination of chromatographic and ORD data can, in principle, be applied to any number of optical isomer components (32), provided there are enough differences in the ORD spectra of the disastereomer components.

III. QUANTITATIVE TLC METHODS OF THE ALKALOID GROUPS

A. Alkaloids of Ornithine and Lysine Origin

Pyrrolidine, pyrrolizidine, pyrazole, pyridine, piperidine, and *quinolizidine alkaloids* belong to this group. For the investigation of these alkaloids several indirect QTLC methods have been worked out, most of them for the determination of *tropine* and *cinchona alkaloids*. These methods are based on elution of the spots followed by fluorometric (e.g., 33) or spectrophotometric (e.g., 34) determination of the alkaloids. According to the method used, the alkaloids can be extracted with chloroform, absolute ethanol, methanol plus 25% ammonia, 0.1 M hydrochloric acid, 0.1 N sulfuric acid, etc. If, instead of silica, magnesium oxide or calcium carbonate and calcium sulfate (35, 36) are used as the adsorbent, the alkaloids can be dissolved together with the adsorbent in 4 N sulfuric acid or 0.1% hydrochloric acid, and in this manner the recovery of alkaloids is complete.

For in situ determinations Dragendorff's reaction was applied for *Leguminosae alkaloids* (37), *tobacco alkaloids* (38), and *tropane alkaloids* (39–41). The in situ reaction is usually followed by densitometry in the reflectance mode at a wavelength, depending on the alkaloids to be measured and the reaction conditions, in the range 400 to 530 nm. In the case of *Leguminosae* alkaloids Faugeras and Paris (37) found the transmittance mode to be more sensitive and precise than reflectance. Naturally, the optimal conditions vary from case to case, and depend largely on the instrument parameters.

Karlsson and Peter (16) determined *Lupinus alkaloids* by fluorodensitometry after a drastic in situ reaction to convert the alkaloids into fluorescent compounds. The alkaloids were determined at an excitation wavelength of 360 nm and an emission wavelength of 400 nm. The blue fluorescence of spots is stable for weeks.

For the investigation of *tobacco alkaloids*, gas chromatography is widely used because of their volatility. However, elution QTLC (42) and densitometry are also used. Massa et al. (43) used fluorescence quenching for the investigation of pseudotropine alkaloids as well.

The tropine alkaloids, which have a condensed pyrrolidine-piperidine skeleton, have only small molar absorptivities at 254 nm. Besides Dragendorff's quantitaiton (e.g., 40) they are determined advantageously by conversion of fluorescent compounds by means of an acidic reagent. Messerschmidt (44) described a method for scopolamine with concentrated sulfuric acid, which was mixed into the mobile phase.

B. Alkaloids of Tyrosine and Phenylalanine Origin

Besides detection with general reagents, a number of other detection methods can be used in this group. Seiler and Weichmann (45) investiaged *Cactus alkaloids* as their dansyl derivatives. The quaternary alkaloids of the *berberine type* are yellow and give strong fluorescence. A fluorodensitometric method was used by Messerschmidt (46) and Hashimoto et al. (47), with 352 and 353 nm wavelengths for excitation and 530 and 510 nm for emission, respectively. For clear selection of the components the two-dimensional technique is sometimes necessary (48).

The *benzophenanthridine alkaloids* can be determined by the elution QTLC method. Balderstone and Dyke (49) determined sanguinarine in *Argemone* oil. They reduced the alkaloid to the dihydro compound to allow extraction with organic solvents. After chromatography the dihydro compound was reconverted by irradiation with long-wavelength UV light. The spots were then eluted with acidic ethanol, and finally the alkaloid was determined spectrophotometrically at 330 nm. Freitag (50) used a dual-wavelength (285 and 365 nm) chromatographic-spectrophotometric method for the determination of *Chelidonium* alkaloids on TLC and HPTLC plates.

Investigating *Hydrastis alkaloids*, Zwaving and de Jong-Havenga (51) compared the direct and elution QTLC methods and found good agreement. In both cases, as with many alkaloids, care must be taken to prevent oxidation.

Among the *bisbenzylisoquinoline alkaloids*, tetrandrine is known for its tumor-inhibiting effect, and tubocurarine as muscle relaxant. In this group, although UV spectrophotometric and indirect paper chromatographic methods are known, QTLC methods have not yet been used.

The *aporphine alkaloids*, including boldine, can be investigated in plant material and pharmaceutical preparation by means of direct QTLC, detecting the spots either with 0.25% 2,6-dichloro- or dibromoquinone chlorimide in methanol (52) or by irradiation with long-wavelength UV light (53).

Among the *Ipecacuanha alkaloids* emetine and cephaeline are frequently investigated in pharmaceutical preparations. They can be detected by indirect QTLC, using spectrophotometry at 283 nm. However, according to Schuyt et al. (54), spectrofluorometry is more sensitive. For emetine, they used an excitation wavelength of 284 nm and measured the emission at 318. According to Massa et al. (43, 55), with a preliminary treatment of the plates with iodine, which oxidizes the alkaloids to yellow fluorescent spots, the alkaloids can be determined by direct spectrofluorometry, measuring the fluorescence at 592 nm after excitation at 358 nm.

Frei and co-workers (17, 56) used a prechromatographic reaction with dansyl chloride for the determination of a number of alkaloids in different pharmaceutical preparations containing, besides emetine and cephaeline, ephedrine, morphine, codeine, and noscapine. Of these compounds, noscapine and codeine do not react with dansyl chloride. A strongly fluorescent dansyl derivative is obtained from morphine as well. They measured the fluorescence at 500–510 nm with excitation at 360 nm. The method is suitable for the analysis of alkaloids in the presence of a 10 to 100-fold excess of other drugs.

For the determination of *opium alkaloids* numerous indirect and direct QTLC methods have been developed. The characteristics of the main opium alkaloids differ significantly, so the detection procedures usually differ from component to component.

Heusser (57) used an elution QTLC method for the determination of morphine as nitrosomorphine, measuring the absorbance of the eluate at 470 nm. Fairbairn and El-Masry (58) separated noscapine and papaverine from the other alkaloids on a thin-layer plate. They eluted the spots of noscapine and papaverine and determined the absorption at 313 and 295 nm for noscapine and 251 and 230 nm for papaverine. Poethke and Kinze (59) separated the principal opimum alkaloids by two-dimensional TLC and determined them spectrophotometrically.

Besides Dragendorff's reaction, several special detection methods are used for the in situ QTLC determination of opium alkaloids. For the determination of morphine in *Papaver* capsules, Paris and Faugeras (61) found detection with sodium nitrate to be most suitable because of the specificity of this reaction compared to other alkaloids. Massa et al. (55) compared different detection methods and determined the optimal conditions for each principal opium alkaloid. The fluorescence quenching at 254 nm and the fluorescence method (excitation at 358 nm, emission at 529 nm) were the optimal ones for papaverine and narceine. Nitrosation was best for detection of morphine, and Dragendorff's reaction was the most suitable for codeine.

Hashiba et al. (62) determined noscapine by fluorodensitometry after heating the plate for 5 h at 120°C. Haefelfinger (63) worked out a fluorodensitometric method for codeine; after separation the plate was sprayed with a mixture of 65% nitric acid and 96% sulfuric acid (4:1), subsequently heated for 10 min at 110°C, and then the fluorescence was measured at 578 nm with excitation at 300 nm. Genest and Belec (64) determined laudanine in opium, using Gibb's reagent for detection; the spots were measured at 465 nm. Röder et al. (65) separated the main opium alkaloids by a two-dimensional method and subsequently measured the absorbance in the reflectance mode. Similar methods were used for codeine and noscapine (66). Jánoș et al. (67) used a reversed phase (RPTLC) plate for the densitometric determination of morphine and quaternary ammonium steroids. Scheidl et al. (68) determined thebaine in *Papaver bracteatum* by densitometry with 1.5% RSD on an HPTLC plate.

C. Alkaloids of Tryptophan Origin

Among the more than 300 known *Strychnos alkaloids*, strychnine and brucine have been most extensively investigated. The majority of QTLC methods are indirect ones (e.g., 69, 70). Massa et al. (43) determined numerous alkaloids, among them strychnine and brucine, densitometrically, using a fluorescence quenching method at 254 nm. Okumura et al. (71) used an FID detector for the determination of a number of alkaloids, including strychnine and brucine, on sintered silica gel or aluminum oxide rods.

The *Rauwolfia alkaloids* (the majority of which are β-carboline-type compounds) occur in plants in great variations, so that their clear separation from each other probably cannot be accomplished by conventional TLC methods with a simple solvent system. Court and co-workers (72, 73) investigated about 300 solvent systems, but none of them was suitable for the clear separation of all the *Rauwolfia alkaloids*. However, they were able to determine the best systems for the separation of reserpine/rescinnamine, ajmalicine/reserpinine, and other difficult cases. They also determined the relationship between structure and retention behavior, which made identification of the components easier. In this group great care must be taken in using QTLC, since these alkaloids are sensitive to light, especially in chloroform and on a dry plate. The analysis must be carried out in the dark or in dimmed light. In this group Dragendorff's reaction modified by Munier and Macheboeuf (5) can be used with success, but fluorescence quenching and the measurement of self-fluorescence usually give better results. A number of elution QTLC methods have been used in this field, e.g., Pötter and Voigt's (74) method. The method developed by Habib and Court (75) is suitable for all alkaloids, including those with a β-carboline skeleton. The method is based on complex formation with iodine and subsequent spectrophotometry. With an elution QTLC method, Manara (76) determined tritium-labeled reserpine by means of a liquid scintillation detector.

The *Rauwolfia alkaloids* can be determined densitometrically. Hartmann and Schnabel (77) determined reserpine and rescinnamine by densitometry at 295 nm. Fluorodensitometric methods are suitable for the analysis of plants (78–80) and other biological materials (81–83) as well. For the fluorodensitometric determination of reserpine (with an excitation wavelength of 392 nm and emission at 540 nm), fluorescence can be obtained by exposure to acetic acid vapor for 24 h (81) or by spraying the plate with 1% *p*-toluenesulfonic acid in glacial acetic acid (82) to develop the fluorescence immediately. Fluorescence can also be developed by exposing the spots to light for at least 24 h to form fluorescent 3-dehydro derivatives (83). Katic et al. (84) determined reserpine and ajmaline in *Rauwolfia vomitara* by an in situ quantitative HPTLC method. They determined ajmalicine by its UV absorption at 292 nm with 5% RSD and reserpine fluorodensitometrically with an excitation wavelength of 380 nm and emission at 450 nm with 3% RSD.

Dombrowski et al. (85) determined the semisynthetic 17-monochloroacetylajmaline and its metabolites by fluorodensitometry after spraying with 6 M nitric acid and acetic acid and gently heating.

For determination of *Mytragina* and *Uncaria alkaloids*, both indirect and in situ QTLC methods have bene known for some time. Shellard and Alam (86–88) compared the densitometric, indirect colorimetric, and spectrophotometric methods and found (in 1968) that the last method proved to be the best one. For densitometry they sprayed the plates with 0.2 M iron(III) chloride in 35% perchloric acid and heated at 120°C for 1.5 h.

The pharmaceutical importance of *Vinca* and *Catharanthus alkaloids* is determined by some alkaloids which originally occured in plants, such as vincamine, vincristine (VCR), and vincaleucoblastine (VLB), and by some semisynthetic and synthetic materials, such as formylleurosine and apovincaminic acid ethyl ester [the active ingredient of Cavinton (Richter Ltd., Budapest, Hungary)].

The binsindol alkaloids isolated from *Catharanthus roseus*, e.g., VLB and VCR, are only minor components in the plant material besides the monomer alkaloids. Adequate quantitative determination of these components requires separation in at least two dimensions, using conventional (non-forced-flow) TLC techniques. Horváth and Iványi (89) developed two different QTLC methods (an indirect spectrophotometric and a densitometric method) for the routine analysis of VLB in *C. roseus* plant material. In the elution method, following a partially selective separation in the first dimension, the spots including VLB (both the investigated and the reference spots) were eluted with a solution containing an internal standard material, VCR. (In the first dimension the VLB spot is far from the VCR spot.) After concentration, the eluates were chromatographed in another solvent system. The VLB and VCR spots were subsequently eluted again, and the absorbances were determined at 289 nm spectrophotometrically. The special internal method eliminated a number of sources of error and ensured satisfactory reliability.

In the densitometric method, the authors employed a two-dimensional chromatographic system on a 20 × 20 cm conventional TLC plate in which the reference spot (VLB) is developed in both dimensions close to the investigated spot but without interfering with it. The results obtained by the two QTLC methods agreed well. However, the direct method is four to five times shorter and requires less manual work. The direct method was also tried on a 10 × 10 cm HPTLC plate, but in this case the quantitative results were less reliable. In the different elution QTLC methods spectrophotometry usually follows chromatography. Karácsony et al. (90) used a nonaqueous titrimetric method after the spot elution. In the in situ method, besides direct spectrophotometry, Massa et al. (43) used Dragendorff's reaction, and Vachnadze (91) used the more specific reaction with Ce(IV) ammonium sulfate. Panas et al. (92) assayed alkaloid-related impurities in vincristine sulfate by a TLC densitometric method, measuring the absorbance at 254 nm.

For the investigation of *physostigmine*, used mostly in ophthalmic solutions, a number of QTLC methods are available. Rogers and Smith (93) determined physotigmine after separation from its degradation products. Mutliwavelength methods were used to separate the useful signal from that of the contamination.

Generally speaking, only higher plants have the ability to biosynthesize alkaloids. The *ergot alkaloids* are the sole exception: they are produced mainly by fungi (*Claviceps purpurea*), but they can also be found in the higher plant "morning glory." Ergot alkaloids can be produced on rye by cultivation or with fermentation.

Elution QTLC methods (e.g., 94, 95) for ergot alkaloids usually begin with an extraction step in order to isolate the alkaloids from the plant material or the fermentation broth. The concentrated (and frequently further purified) extract is applied to the TLC plate, and after chromatography, direct spectrophotometry with van Urk reagent follows. Sometimes indirect fluorometric analysis is used (94).

In the densitometric methods most analysts measure the self-fluorescence of the alkaloids, where possible. Different authors have used the following excitaiton and emission wavelengths: for ergometrine, 358 and 490 nm (95) or 313 and 420 nm (96); for LSD, 330 and 410 nm (97); and for ergotamine, 313 and 420 nm (96) or 325 and 445 nm (98, 99). Frijns (96) impregnated the developed plates with liquid paraffin in diethyl ether (1:9) to stabilize the fluorescence. Horváth (100) used a 15% paraffin solution in carbon tetrachloride for this purpose. The paraffin film enhances the fluorescence by an order of magnitude.

A critical problem in TLC of the lysergic acid derivative ergot alkaloids is separation of the ergotoxin components from each other, especially α- and β-ergocrystine and ergocornine. Only two chromatographic systems comply with these requirements. Reichelt (101) used a partition chromatographic system with diisopropyl ether-tetrahydrofuran-toluene-diethylamine (70:15:15:0.1) on precoated silica gel plates impregnated with a mixture of 18 ml formamide, 0.6 ml 25% ammonia, and 81.4 ml acetone. Szepesi et al. (102) used a more reproducible adsorption chromatographic

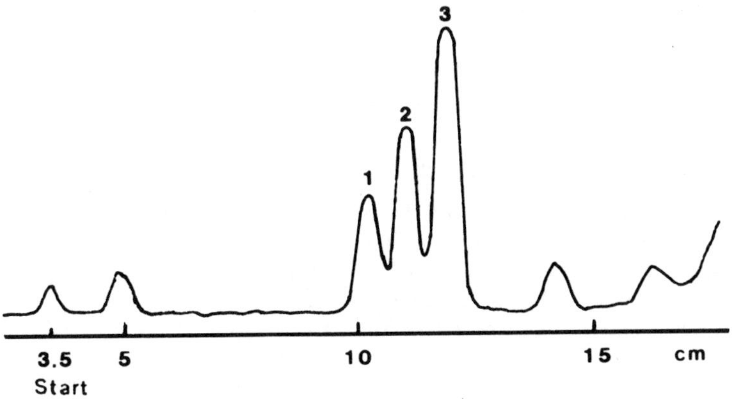

Figure 3 Densitogram of an ergot extract sample. 1, β-ergocryptine; 2, α-ergocryptine; 3, ergocornine. OPTLC conditions: cushion pressure, 1.0 MPa; flow velocity, 0.08 ml/min; time of development, 35 min; scanning, Opton KM 3 chromatograph-spectrophotometer; fluorescence measurement, excitation at 366 nm, emission at 456 nm.

system for the same purpose, at the same time solving the separation problem of dihydroergotoxine alkaloids as well. In this system the stationary phase is silica gel G UV_{254} (on Macherey Nagel precoated plates) and the mobile phase is acetone-0.1 M ammonium carbonate-ethanol (32.5:67.5:1). The only drawback of this system is that the development is slow, 3h is necessary for 15 cm migration. I have modified this sytem (see Fig. 3) and applied the method in an overpressurized chamber, where optimum separation was obtained with a development lasting for only 35 min.

In fluorodensitometric determination of the dihydro ergot alkaloids the emitted light is not in the visible range. However, the alkaloids can be converted on the plate by UV irradiation (usually 1 h at 254 nm) into the normally fluorescent components (e.g., 103). Prosek et al. (99) used a fluorodensitometric method for these alkaloids without converting them. They determined the emitted light at 250–320 nm with excitation at 230 nm. This method was better than the determination of absorbance at 280 nm. Instead of direct densitometry or fluorodensitometry, some authors (e.g., 104, 105) prefer the van Urk reaction and subsequent densitometry. For example, Genest (104), investigating ergot alkaloids in morning glory, sprayed the developed plate with van Urk reagent. After spraying, the plates were stored in a dark place for 20 min for stabilization sprayed with 1% sodium nitrate, and the absorbances were measured at 570 nm.

D. Alkaloids of Anthranilic Acid Origin

The *Cinchona alkaloids*, including those with a quinoline skeleton, belong to this group. Quinine and quinidine are used as drugs and are also important materials in the food industry. Cinchonine and cinchonidine are also frequently investigated alkaloids.

Several indirect QTLC methods are known, based on spectrophotometric (e.g., 106, 107) or fluorometric (108, 109) determination. The spectrophotometric determinations are carried out at wavelengths in the long-wavelength UV range; in the fluorometric methods the excitation wavelength is 350 nm and the emission wavelength is 450–456 nm for quinine and quinidine. Studying the optimum conditions for in situ QTLC methods, Massa et al. (43, 110) found absorption maxima for quinine and quinidine at 330 nm and for cinchonine and cinchonidine at 288 nm, after spraying with ethanol-concentrated sulfuric acid (9:1). The detection limit was about 100 ng. By means of fluorometry, using 1% sulfuric acid in ethanol as a spray reagent, the detection limit is 1 ng for quinine and quinidine at an excitation wavelength of 365 nm and measurement at 450 nm, and 5 ng for cinchonine and cinchonidine with excitation at 313 nm and emission at 410 nm. Röder et al (111) used diethyl ether-concentrated sulfuric acid (95:5) as an immersion reagent and measured the fluorescence of quinine and quinidine at 460 nm with excitation at 365 nm. Under these conditions cinchonine and cinchonidine did not interfere. The latter compound can also be determined fluorodensitometrically without interference from quinine and quindine by excitation at 313 nm and measurement of the emission at 390 nm. Ebel and Herold (112) used this method for the determination of quinine without complete separation. Okumura et al. (71) used the glass rod separation method with an FID detector for the determination of quinine.

E. Alkaloids of Histidine Origin

The imidazole alkaloids are used mostly in ophthalmic preparations. Massa et al. (43) used an in situ chromatographic method with Dragendorff's reagent for the determination of pilocarpine, isopilocarpine, and pilocarpic acid in pharmaceutical preparations.

F. Steroidal Alkaloids

Steroidal alkaloids occur mainly in *Solanum*, *Veratrum*, and *Holarrhena* species. Solanin occurring in potato has been known since 1820. For 20 years indirect spectrophotometric QTLC methods have been used (113, 114) for the detemrination of *Solanum glycoalkaloids*.

For the direct QTLC determination of *Solanum* alkaloids, Fayez and Saleh (115) used a 5% solution of phosphomolybdenic acid in ethanol containing 4 ml sulfuric acid per 100 ml. The plates were heated for 10 min at 110°C. Cadle et al. (116) used antimony(III) trichloride reagent in chloroform and heating at 150°C for 4 min for the determination of glycoalkaloids in potatoes.

For densitometric purposes Dragendorff's reaction is also suitable in this group (43). Jellema et al. (117) used an optical brightner, Blankophor BA 267% (Bayer, Leverkusen, FRG), for the fluorodensitometric determination of potato glycoalkaloids by TLC. The excitation wavelength was 300 nm, and the measurement was carried out at 450 nm.

G. Xanthine Alkaloids

The xanthine alkaloids cannot be detected directly by Dragendorff's reaction and iodoplatinate reagent except for compounds with an amino group in the side chain. These alkaloids, which include caffeine, theobromine, and theophylline, can be detected spectrophotometrically at their absorption maximum, 272 nm. A number of indirect QTLC methods are used for the determination of caffeine (e.g., 118, 119) and theophylline (e.g., 120) in biological material. Ebel and Herold (121–123) developed methods for the investigation of caffeine in pharmaceutical preparations. They used phenacetin as an internal standard and two-dimensional TLC for the separation.

Some authors (124–127) have discussed the direct densitometric determination of caffeine in plant materials by measuring the quenching of UV light at 254 nm on fluorescent plates. Malingre and Batterman (128) found this method to be less sensitive than densitometry at the absorption maximum of 273 nm. Laub and Zimmer (129) used absorption measurements at 273 nm in the reflectance mode for rapid TLC determination of theobromine in chocolate.

H. Miscellaneous Groups

Most of the work on TLC of *diterpene alkaloids* concerns the alkaloid aconitine. For its elution QTLC determination it is useful to carry out the chromatography on magnesium oxide (130) and subsequently to dissolve the adsorbent with the alkaloid spot, because of its strong adsorption to the silica. Massa et al. (43) used an in situ QTLC method with Dragendorff's reaction and detection at 400 nm.

For the determination of *colchicine alkaloids* only some indirect QTLC methods are known so far (e.g., 131); the components are detected by UV spectrophotometry.

Traditionally, some *compounds of the amide type* have been included among the alkaloids; these are mainly capsaicine and its analogs, which have been investigated primarily in the food industry. These compounds can be detected well with general alkaloid reagents. Suzuki et al. (132) reported an effective separation of capsaicine and related alkaloids with a reversed phase HPTLC system. They used ammonia vapor and then 0.1% 2,6-dichloroquinone-4-chloroimide in methanol for detection, and by in situ densitometric measurement at 610 nm they could determine these components with much better results than were obtained with the earlier TLC, GLC, and HPLC methods.

The QTLC determination of *quaternary ammonium compounds* presents some special problems, in both the separation and the detection of these highly polar compounds. Besides normal phase chromatography, usually with highly polar solvent systems, partition (133–135) and ion pair chromatography (136–141) have been more frequently used for the separation of these compounds than for other compounds. The quaternary ammonium compounds can be readily detected with iodobismutate and iodoplatinate reagents, and for compounds that do not have a chromophore these reactions usually give the best, and sometimes the only suitable; alternative for QTLC determination.

Some pharmaceutically important and very effective muscle relaxant steroid compounds belong to this group. Figure 4 shows a densitomgram for the separation of pipecuronium bromide [2β,16β-bis(4'-dimethylpiperazino)-3α,17β-diacetoxy-5α-androstane dibromide] in an artificial mixture with its analogs. The quantity of analogous components is 1% each, realtive to the main component. In this densitogram a correlation can be observed between the chemical structure and R_f values. The bis-quaternary, monoquaternary, and bis-tertiary compounds are separated completely in three bands in the chromatogram. The separation of certain component pairs in the bands is not complete but is sufficeint for reliable quantitation.

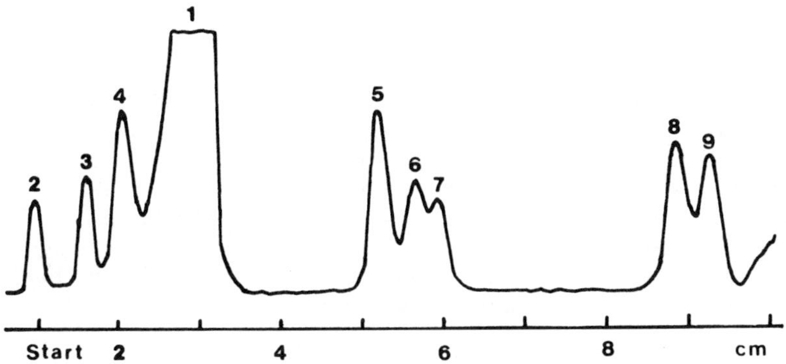

Figure 4 Densitogram of pipecuronium bromide and related compounds. 1, pipecuronium bromide; 2-4, related bisquaternary compounds; 5-7, related monoquaternary (monotertiary) compounds; 8-9, related bistertiary compounds. Layer, HPTLC silica gel 60; mobile phase, 1.4 g ammonium carbonate-20 ml 25% ammonia solution-80 ml methanol; densitometry after Dragendorff's reaction, absorbance measurement in reflectance mode at 525 nm; instrument, Opton KM 3 chromatograph-spectrophotometer.

IV. CONCLUSIONS

At the time of the preparation of this chapter nearly 400 QTLC methods dealing with alkaloids could be found in the chemical literature. Most of them were developed, or at least applicable, for industrial purposes. This is one of the fields, among the numerous industrial applications of liquid chromatography, in which QTLC compares favorably with HPLC. In industrial analytical practice chemists are often satisfied with quantitative data generated for the purpose of rapid orientation. This allows the use of QTLC procedures, which are not optimized in every respect. It is feasible that the number of methods actually used far exceeds the number of methods published in journals. A substantial portion of the published articles on QTLC still report indirect methods, although more recently the in situ instrumental methods have gained importance.

The majority of alkaloids can be determined quantitatively by use of a general reagent, such as Dragendorff's reagent, to name the most important one. In many instances, however, the inherent spectrophotometric or fluorometric character of the compounds or a more specific reagent gives better results.

Since many alkaloids are generally known and are easy to determine, they are frequently used as model compounds to demonstrate the effectiveness of technical modificaitons of TLC, so that there can scarcely be any new step in QTLC development wihtout its demonstration in the field of alkaloids as well. However, it appears that today the simple reflectance densitometric and "near-side" fluorodensitometric methods constitute the bulk of applied instrumental QTLC procedures. The novel technical approaches devised to increase the signal-to-noise ratio and provide better resolution and the new detection techniques have still been utilized only in a modest number of cases.

REFERENCES

1. I. M. Hais and K. Macek. *Handbuch der Papierchromatographie*, VEB Gustav Fisher Verlag, Jena, 1958.
2. G. Hauck. Dtsch. Apoth. Ztg. *105*, 209 (1965).
3. F. Pellerin, D. Dumitrescu-Mancheron, and C. Chabrelle. Bull. Soc. Chim. Fr. 123 (1973).
4. R. Munier. Bull. Soc. Chim. Biol. *35*, 1225 (1953).
5. R. Munier and M. Macheboeuf. Bull. Soc. Chim. Biol. *31*, 1144 (1949).
6. H. Thies and F. W. Reuther. Naturwissenschaften *41*, 230 (1954).
7. D. Vágújfalvi. Planta Med. *13*, 79 (1965).
8. H. Trabert. Pharm. Zentralhalle *93*, 463 (1954).
9. W. W. Fike. Anal. Chem. *38*, 1697 (1966).
10. A. Puech, M. Jakob, and D. Gaudy. J. Chromatogr. *68*, 161 (1972).
11. J. C. Touchstone and M. F. Dobbins. *Practice of Thin-Layer Chromatography*, p. 165, Wiley, New York, 1978.
12. L. J. Glunz. In: *Thin-Layer Chromatography: Quantitative Environmental and Clinical Applications*, J. C. Touchstone and D. Rogers (eds.), p. 85, Wiley, New York, 1980.
13. R. Segura and A. M. Gotto, Jr. J. Chromatogr. *99*, 643 (1974).

14. P. R. Bochoff, B. J. Kopkins, and V. Pretorius. J. Chromatogr. *126*, 35 (1976).

15. L. Zhou, S. Shanfield, and A. Zlatkis. J. Chromatogr. *239*, 259 (1982).

16. E. M. Karlsson and H. W. Peter. J. Chromatogr. *155*, 218 (1978).

17. R. W. Frei, W. Santi, and M. Thomas. J. Chromatogr. *116*, 365 (1976).

18. K. Macek. In: *Pharmaceutical Applications of Thin-Layer and Paper Chromatography*, K. Macek (ed.), p. 431, Elsevier, Amsterdam, 1972.

19. F. Santavy. In: *Thin-Layer Chromatography*, E. Stahl (ed.), p. 421, Springer-Verlag, Berlin, 1969.

20. E. Tyihak and D. Vagujfalvi. In: *Progress in Thin-Layer Chromatography and Related Methods*, vol. 3, A. Niederwieser and G. Pataki (eds.), p. 71, Ann Arbor, London, 1972.

21. J. G. Kirchner. In: *Techniques of Chemistry*, Vol. 14, *Thin-Layer Chromatography*, A. Weissberger (ed.), p. 386, Wiley, New York, 1978.

22. D. Vagujfalvi. In: *Pocket Book of Thin-Layer Chromatography*, E. Tyihak (ed.), p. 451, Müszaki Könyvkiadó, Budapest, 1979.

23. A. Baerheim Svendsen and R. Verpoorte. In: *Chromatography of Alkaloids*, Part A, *Thin Layer Chromatography*, Journal of Chromatography Library, vol. 23A, Elsevier, Amsterdam, 1983.

24. A. Der Marderosian and E. Songu. In: *Densitometry in Thin-Layer Chromatography*, J. C. Touchstone and J. Sherma (eds.), p. 135, Wiley, New York, 1979.

25. E. Tyihak, E. Mincsovics, and H. Kalasz. J. Chromatogr. *174*, 75 (1979).

26. P. Horváth, G. Szepesi, M. Hoznek, Z. Vegh, and E. Mincsovics. In: *Proceedings of the 2nd International Symposium on Instrumental High Performance Thin-Layer Chromatography*, p. 295, Bad Dürkheim, Federal Republic of Germany, 1982.

27. H. F. Hauck and W. Jost. J. Chromatogr. *262*, 113 (1983).

28. H. Gulyás and E. Mincsovics. Personal communication (1984).

29. P. Horváth, G. Szepesi, and A. Kassai. Planta Med. *33*, 407 (1978).

30. D. Eskes. J. Chromatogr. *152*, 589 (1978).

31. P. Horváth, L. Terdy, and F. Sáry. Magy. Kém. Foly. *84*, 449 (1978); Chem. Abstr. *90*, 66254a.

32. P. Horváth: Simultaneous determination of optical isomers by a combination of chromatographic and O.R.D. methods. In: *Euroanalysis V*, Cracow, 1984.

33. E. Smith, S. Barkan, B. Ross, M. Maienthal, and J. Levine. J. Pharm. Sci. *62*, 1151 (1973).

34. S. M. Karawya and A. M. Diab. J. Pharm. Sci. *66*, 1317 (1977).

35. A. Suszko-Purzycka and W. Trzebny. J. Chromatogr. *16*, 239 (1964).

36. F. Wartmann-Hafner. Pharm. Acta Helv. *41*, 406 (1966).

37. G. Faugeras and M. Paris. Bull. Soc. Chim. Fr. 109 (1973).

38. A. Jurzysta. Pamiet. Pulawski *62*, 159 (1975).

39. Z. P. Kostennikova and V. B. Chichiro. Farmatsiya (Moscow) *18*, 39 (1969); Chem. Abstr. *71*, 128782q (1969).

40. B. L. Wu Chu, E. S. Milka, M. J. Solomon, and F. A. Crane. J. Pharm. Sci. *58*, 1073 (1969).

41. M. S. Shipalov, V. E. Chichiro, and Z. P. Kostennikova. Prikl. Biokhim. Mikrobiol. 5, 502 (1969); Chem. Abstr. *71*, 116567k (1969).

42. R. M. Navari. J. Chem. Educ. *51*, 748 (1974).

43. V. Massa, F. Gal, and P. Susplugas. *International Symposium on Chromatography and Electrophoresis*, 6th Lecture Paper, p. 470 (1970).

44. W. Messerschmidt. Dtsch. Apoth. Ztg. *109*, 199 (1969).

45. N. Seiler and M. Weichmann. Z. Anal. Chem. *220*, 109 (1966).

46. W. Messerschmidt. J. Chromatogr. *39*, 90 (1969).

47. Y. Hashimoto, K. Ando, and M. Mizuno. Shoyakugaku Zasshi *30*, 127 (1976); Chem. Abstr. *87*, 29066h (1977).

48. T. Hattori, M. Inoue, and M. Hayakawa. Yakugaku Zasshi *97*, 1263 (1977); Chem. Abstr. *88*, 126386r (1978).

49. P. Balderstone and S. F. Dyke. J. Chromatogr. *132*, 359 (1977).

50. W. E. Freitag. Planta Med. *40*, 278 (1980).

51. J. H. Zwaving and E. H. J. de Jong-Havenga. Pharm. Weekbl. *107*, 137 (1972).

52. N. Di Renzo. Boll. Chim. Farm. *111*, 450 (1972).

53. L. Fleischmann, G. Maffi, and P. G. Mezzenzanica. Boll. Chim. Farm. *114*, 534 (1975).

54. C. Schuyt, G. M. J. Beiersbergen van Henegouwen, and K. W. Gerritsma. Analyst *102*, 298 (1977).

55. W. Massa, F. Gal, P. Susplugas, and G. Maestre. Trav. Soc. Pharm. Montpellier *30*, 301 (1970).

56. F. Nachtmann, H. Spitzey, and R. W. Frei. Anal. Chim. Acta *76*, 57 (1975).

57. D. Heusser. J. Chromatogr. *33*, 400 (1968).

58. J. W. Fairbairn and S. El-Masry. J. Pharm. Pharmacol. *195*, 935 (1967).

59. W. Poethke and W. Kinze. Arch. Pharm. (Weinheim) *297*, 593 (1964).

60. M. von Schantz. In: *Thin-Layer Chromatography*, G. B. Martini-Bettolo (ed.), p. 122, Elsevier, Amsterdam, 1964.

61. R. R. Paris and G. Faugeras. Ann. Pharm. Fr. *24*, 613 (1966).

62. S. Hashiba, M. Tatsusawa, and A. Ejima. Bunseki Kagaku (Jpn. Anal.) *26*, 804 (1977); Chem. Abstr. *88*, 126387s (1978).

63. P. Haefelfinger. J. Chromatogr. *124*, 351 (1976).

64. K. Genest and G. Belec. Can. J. Pharm. Sci. *2*, 44 (1967).

65. K. Röder, E. Eich, and E. Mutschler. Arch. Pharm. (Weinheim) *304*, 297 (1971).

66. W. Schlemmer and E. Kammerl. J. Chromatogr. *82*, 143 (1973).

67. É. János, T. Cserháti, and E. Thihák. J. High Resol. Chromatogr. *5*, 634 (1982).

68. K. Scheidl, F. Wurst, Th. Prey, and E. Bancher. Z. Anal. Chem. *308*, 428 (1981).

69. D. D. Datta, P. C. Bose, and D. Ghosh. East. Pharm. *15*, 41 (1972).

70. M. S. Karawya, M. S. Hifnawy, and H. S. Dahawi. J. Assoc. Off. Anal. Chem. *58*, 85 (1975).

71. T. Okumura, T. Kadono, and A. Isoo. J. Chromatogr. *108*, 329 (1975).

72. W. E. Court and M. S. Habib. J. Chromatogr. *80*, 101 (1973).

73. W. E. Court and P. Timmins. Planta Med. *27*, 319 (1975).

74. H. Pötter and R. Voigt. Pharmazie *22*, 198 (1967).

75. M. S. Habib and W. E. Court. Can. J. Pharm. Sci. *8*, 81 (1973).

76. L. Manara. Eur. J. Pharmacol. *2*, 136 (1967).

77. V. Hartmann and G. Schnabel. Pharm. Ind. *37*, 451 (1975).

78. E. Kammerl and E. Mutschler. Pharm. Ztg. *118*, 1905 (1973).

79. D. Heusser. Planta Med. *12*, 237 (1964).

80. F. Hammer and F. Kaiser. Planta Med. *21*, 5 (1972).

81. S. L. Tripp, E. Williams, W. E. Wagner, Jr., and G. Lukas. Life Sci. *16*, 1167 (1975).

82. R. A. Sams and R. Huffman. J. Chromatogr. *161*, 410 (1978).

83. J. M. G. J. Frijns. Pharm. Weekbl. *106*, 605 (1971).

84. M. Katic, E. Kucan, M. Prosek, and M. Bano. J. High Resol. Chromatogr. *3*, 149 (1980).

85. L. J. Dombrowski, A. V. Crain, R. S. Browning, and E. L. Pratt. J. Pharm. Sci. *64*, 643 (1975).

86. E. J. Shellard and M. Z. Alam. J. Chromàtogr. *33*, 347 (1968).

87. E. J. Shellard and M. Z. Alam. J. Chromatogr. *35*, 72 (1968).

88. E. J. Shellard. In: *Quantitative Paper- and Thin-Layer Chromatography*, E. J. Shellard (ed.), p. 51, Academic Press, London, 1968.

89. P. Horváth and G. Ivanyi. Acta Pharm. Hung. *52*, 150 (1982).

90. E. M. Karácsony, I. Gyenes, and Cs. Lörincz. Acta Pharm. Hung. *35*, 280 (1965).

91. V. Yu. Vachnadze. Izv. Akad. Nauk. Gruz. SSR Ser Khim. *1*, 28 (1975).

92. J. M. Panas, M. F. Etcheverry, J. M. Yzerman, and G. Ledouble. Ann. Pharm. Fr. *37*, 49 (1979).

93. A. R. Rogers and G. Smith. J. Chromatogr. *87*, 125 (1973).

94. M. Klavehn, H. Rochelmeyer, and J. Seyfried. Dtsch. Apoth. Ztg. *101*, 75 (1961).

95. Ya. Petrova, T. Tomova, and L. Filipova. Farmatsiya (Sofia) *22*, 9 (1972).

96. J. M. G. J. Frijns. Pharm. Weekbl. *106*, 865 (1971).

97. T. Niwaguchi and T. Inove. J. Chromatogr. *59*, 127 (1971).

98. M. Prosek, E. Kucan, M. Katic, and M. Bano. Chromatographia *9*, 273 (1976).

99. M. Prosek, E. Kucan, M. Katic, and M. Bano. Chromatographia *9*, 325 (1976).

100. P. Horváth. Unpublished data (1977).

101. J. Reichelt, Cesk. Farm. *25*, 93 (1976).

102. G. Szepesi, J.Molnár, and Sz. Nyiredy. Z. Anal. Chem. *294*, 47 (1979).

103. C. J. G. A. Bos and J. M. G. J. Frijns. Pharm. Weekbl. *107*, 111 (1972).

104. K. Genest. J. Chromatogr. *19*, 531 (1965).

105. M. Vanhaelen and R. Vanhaelen-Fastre. J. Chromatogr. *72*, 139 (1972).

106. G. Baerwald and J. Prucha. Monatsschr. Brau. *26*, 190 (1973).

107. S. M. Karawya and A. M. Diab. J. Pharm. Sci. *66*, 1317 (1977).

108. E. Ragazzi and G. Veronese. Michrochim. Acta 966 (1965).

109. E. Smith, S. Barkan, B. Ross, M. Maienthal, and J. Levine. J. Pharm. Sci. *62*, 1151 (1973).

110. V. Massa, P. Susplugas, and R. Taillade. Trav. Soc. Pharm. Montpellier *32*, 141 (1974).

111. K. Röder, E. Eich, and E. Mutschler. Pharm. Ztg. *115*, 1430 (1970).

112. S. Ebel and G. Herold. Z. Anal. Chem. *266*, 281 (1973).

113. N. A. Valocich and L. Nad. Med. Prom. SSSR *20*, 44 (1966); Chem. Abstr. *64*, 20168d (1966).

114. H. Roensch and K. Schreiber. J. Chromatogr. *30*, 149 (1967).

115. M. B. E. Fayez and A. A. Saleh. Z. Anal. Chem. *246*, 380 (1969).

116. L. S. Cadle, D. A. Stelzig, K. L. Harper, and R. J. Young. J. Agric. Food Chem. *26*, 1453 (1978).

117. R. Jellema, E. T. Elema, and Th. Malingre. J. Chromatogr. *210*, 121 (1981).

118. M. Amin and W. Sepp. J. Chromatogr. *118*, 225 (1976).

119. M. Reichert. J. Chromatogr. *146*, 175 (1978).

120. R. N. Gupta, F. Eng, and M. Stefanec. Clin. Biochem. *11*, 42 (1978).

121. S. Ebel and G. Herold. Chromatographia *8*, 35 (1975).

122. S. Ebel and G. Herold. Chromatographia *8*, 569 (1975).

123. S. Ebel and G. Herold. Chromatographia *9*, 41 (1976).

124. V. Massa, F. Gal, P. Susplugas and G. Maestre. Trav. Soc. Pharm. Montpellier *32*, 141 (1974).

125. L. R. Sellier and V. Torre. Cienc. Ind. Farm. *4*, 146 (1972); Chem. Abstr. *82*, 160280u (1975).

126. G. Dertinger and H. Scholz. Pharm. Ind. *34*, 114 (1972).

127. J. Sherma and M. Beim. J. High Resol. Chroamtogr. *1*, 309 (1978).

128. T. M. Malingre and S. Batterman. Pharm. Weekbl. *112*, 1305 (1977).

129. E. Laub and M. Zimmer. Chemie *33*, 117 (1979).

130. A. Hiermann, Zentralbl. Pharm. Pharmakother. Laboratoriumsdiagn. *113*, 1247 (1974).

131. H. Potesilova, J. Santavy, A. El-Hamidi, and F. Santavy. Collect. Czech. Chem. Commun. *34*, 3540 (1969).

132. T. Suzuki, T. Kawasa, and K. Iwai. J. Chromatogr. *198*, 217 (1980).

133. A. Bonati and M. Bacchini. Fitoterapia *37*, 24 (1966).

134. A. Noirfalise and G. Mees. J. Chromatogr. *31*, 61 (1961).

135. G. H. Jolliffe and E. J. Shellard. J. Chromatogr. *48*, 125 (1970).

136. F. Wartmann-Hafner. Pharm. Acta Helv. *41*, 406 (1966).

137. H. Potesilova, J. Hrbek, Jr., and F. Santavy. Collect. Czech. Chem. Commun. *32*, 141 (1967).

138. N. H. Eitel, S. L. Wallace, and B. Omokoku. Biochem. Med. *4*, 181 (1970).

139. K. F. Ahrend and D. Tiess. Wiss. Z. Univ. Rostock Math. Naturwiss. Reihe *22*, 951 (1973).

140. R. J. Armstrong, N.Z. J. Sci. *17*, 15 (1974).

141. F. W. Grant, J. Chromatogr. *116*, 230 (1976).

9

Monitoring Fermentation Processes and Analysis of Some Antibiotics

FRANZ KREUZIG *Biochemie GmbH, Kundl, Austria*

I. INTRODUCTION

Many substances related to food, health, and other areas of daily life are produced by fermentation. Biotechnology is a rapidly growing branch of industry driven by new genetic techniques and progress in fermenter design and equipment, computer applications, and fast analytical methods.

Most of the antibiotics available are now produced by fermentation, because syntheses are often too expensive and complicated. Fermentations for producing antibiotics require fast analytical methods for the determination of different substances, and the antibiotic itself in the fermentation liquid. Considering the fact that the costs of these liquids in industrial fermenters are very high, the need for fast and reliable analytical methods is obvious.

On-line methods would be optimal as they provide automatic control of the process. But technical problems in automated sampling (of broth and filtrate), automated weighing of the broth, and further treatment of the sample have not yet been solved. Analytical determination of the final product, by-products, and other components of the fermentation liquid, and of substances added during the process, is frequently done by chromatographic methods. This is necessary because the substances to be analyzed are usually present in low concentrations, and removal of the interfering biological matrix is often impossible by any other technique. Therefore quantitative high-performance TLC has been introduced (1,2).

II. INSTRUMENTATION

A. Plates

Silica gel plates for high-performance TLC (HPTLC) have been made available by E. Merck and Macherey & Nagel. One fact should be mentioned: in order to have a hard layer the manufacturers are adding organic binders which absorb to a certain extent in the ultraviolet (UV). According to the composition of the solvent, this binder may migrate in the layer.

B. Application

The volumes of both standards and samples are 250 nl. This is a compromise between detectability, accuracy of dosage, and capacity of the plate for many samples. The application scheme is that

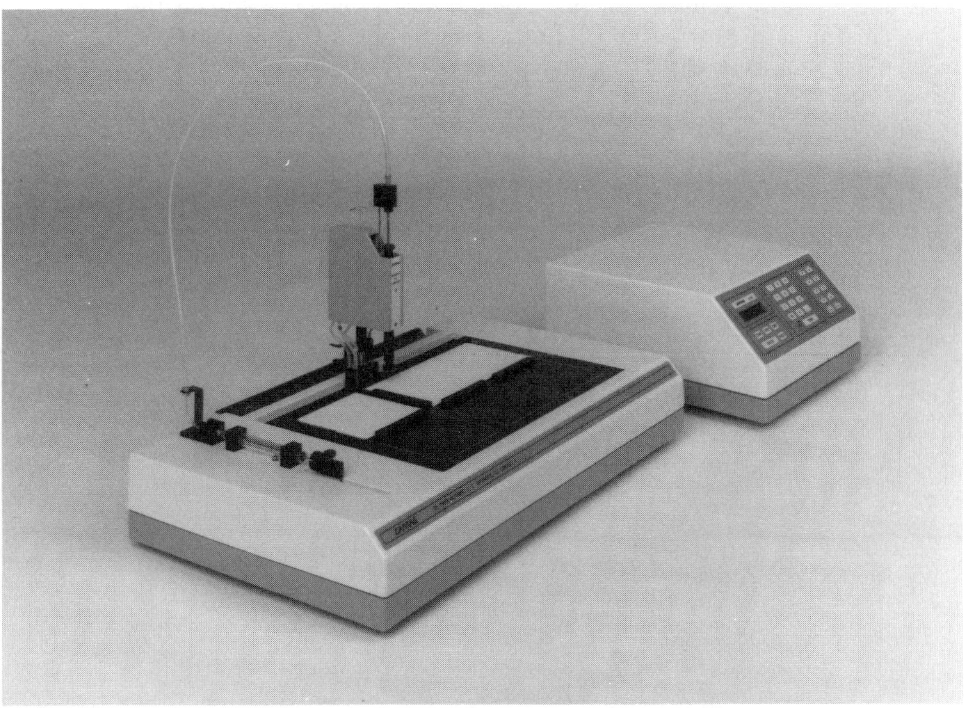

Figure 1 CAMAC TLC applicator II.

of the data pair technique (3). Here 1–12 are samples and a–c are standards.

1 2 3 a 4 5 6 b 7 8 9 c 10 11 12 1 2 3 a 4 5 6 b 7 8 9 c 10 11 12

According to the size of the resulting spots, it may be necessary to reduce the number of samples on one plate. The following instruments are in use. For manual application of hexane, chloroform, methylene chloride, ethyl acetate, and butyl acetate solutions, custom-made Pt-Ir capillaries are used. For water, ethanol, and methanol solutions we use the Nano-applicator from CAMAG.

In addition to the above instruments the CAMAG TLC-applicator II (see Figure 1) is used. It takes about 16 min to apply twice 12 samples and 3 standards according to the scheme mentioned above. The following optimal apparatus conditions have been found: delivery speed, 500 nl/s; air segment, 100 nl; excess volume, 3000 nl; predosage volume, 500 nl; rinsing volume, 2100 nl; spotting volume, 250 nl. Degassed methanol has been the rinsing and replacement liquid. Sample vials from CAMAG with aluminum caps and from Chromglas (D-8229 Haag, FRG) with screw caps are used.

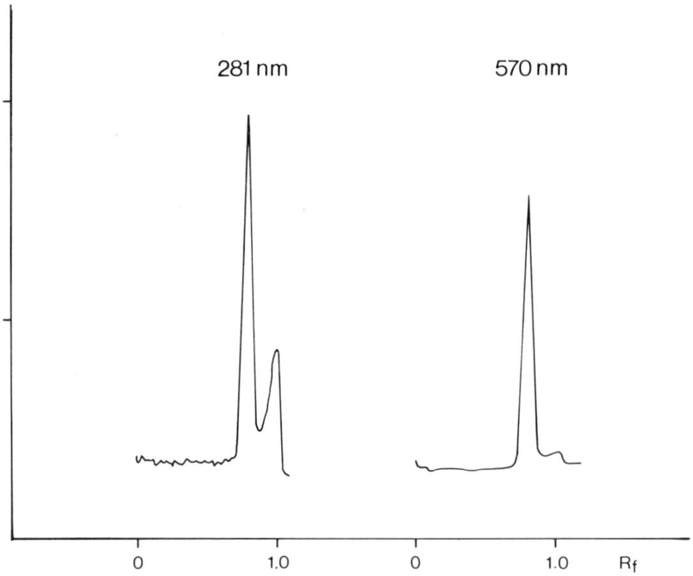

Figure 2 Derivatization of gramicidin.

C. Separation

Although conventional linear chromatography in saturated chambers (Micro-Chamber, E. Merck, No. 11622, and Twin Trough Chamber, CAMAG) meets all the requirements of our HPTLC separations, it should be noted that for special separation problems circular, anticircular, or overpressurized HPTLC can be advantageous.

D. Drying

The plates are dried in an upright position in an oven. A uniform temperature should be provided in the oven.

E. Derivatization

A typical example of the need for derivatization is the determination of the polypeptide antibiotic gramicidin. Scanning at the wavelength 281 nm, where the absorption shows a maximum, results in an additional peak close to the gramicidin peak due to the binder of the layer. This can be avoided by derivatization with 4-dimethylaminobenzaldehyde and scanning at 570 nm (4), as illustrated by Figure 2.

Derivatization can be performed in different ways. In special cases it is possible to include the reagent in the layer; after the separation the plate is heated, giving the spot a characteristic color. Another method is derivatization with gas (NH_3, HCl, HBr), generating fluorescence. Nitrous gases can intensify the UV-absorption of aromatic compounds. With this method it has been possible to show that phenylacetic acid, penicilloic acid, and penicillin G, all with relatively low UV absorbances, can be scanned at lower concentrations on HPTLC plates (4, 5). One of the usual postchromatographic derivatization methods is dipping the plate into the reagent. This is possible when polarities of samples and reagent are different so that the spots are neither dissolved nor diffused. It should be mentioned that the reagent can be contaminated with inpurities from the plates, so the reagent must be discarded frequently. It may also happen that contaminants of the sample are chromatographed close to the spot to be measured. Furthermore, gradient formation by dipping reagents has been reported (6).

In order to avoid such limitations, it is advisable to spray the reagent onto the plate. This procedure can be performed in a reproducible manner by means of a spraying device (A. Paar KG, Graz, Austria; Figure 3) (7–9). The plate is moved horizontally from right to left while the spraying vessel simultaneously moves up and down. This technique is also useful for circular and anti-

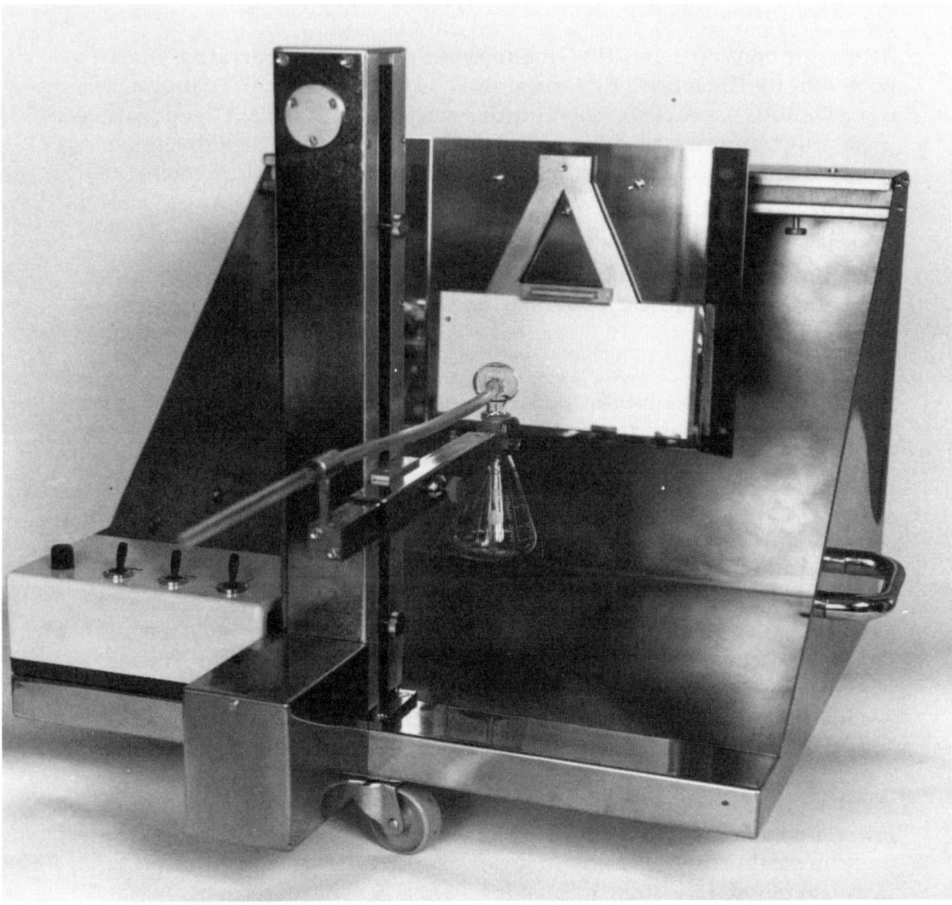

Figure 3 Automated spraying device.

circular chromatograms, because dipping will cause one-way deformation of the spots.

The difference in peak shape between dipping and spraying can be seen in Figure 4. Gramicidin samples 1 to 3 were spotted twice and chromatographed. After separation and drying the plate was cut; one part was sprayed and the other was dipped. It can be seen that by-products of the sample were chromatographed too close to the gramicidin spot when the plate was dipped into the reagent. The sloping arrows indicate this phenomenon.

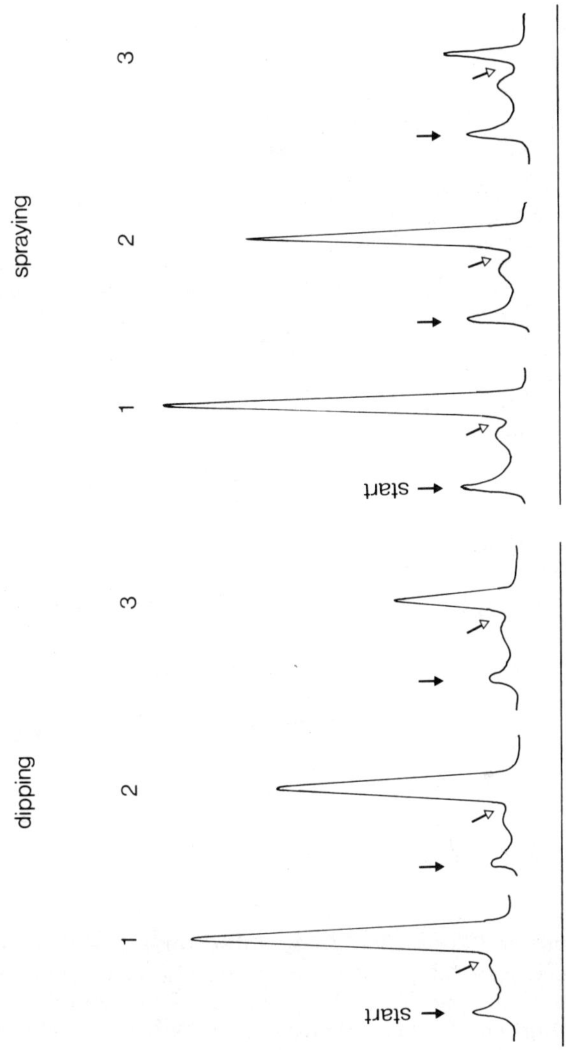

Figure 4 Peak shape depending on dipping and spraying.

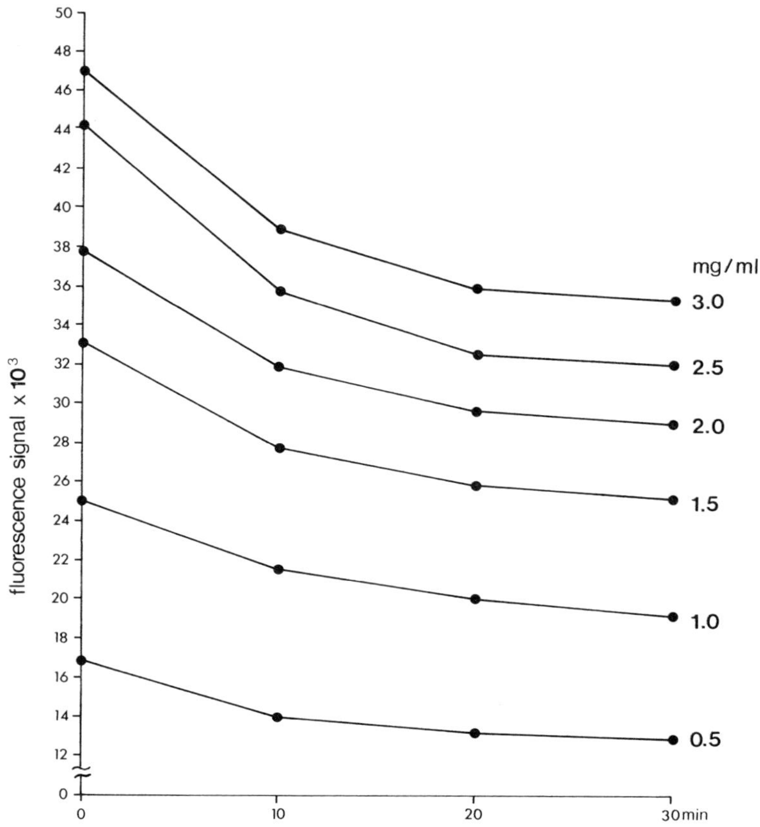

Figure 5 Stability of methionine spots, derivatized with Fluram.

The derivatized spot should be stable for at least 30 min. In special cases it may happen that a stable signal is avaialble a certain time after the reaction. An example in the determination of methionine (Figure 5; derivatization with Fluram). The resulting signal decreases by about 24% within a half-hcur, after which sufficient stability is reached (4).

For heating the plate after soaking with the reagent, a CAMAG TLC plate heater II has been used.

F. Measurement and Evaluation

A high scanning speed is necessary in order to shorten the analysis time, and, of course, automated acquisition and evaluation of

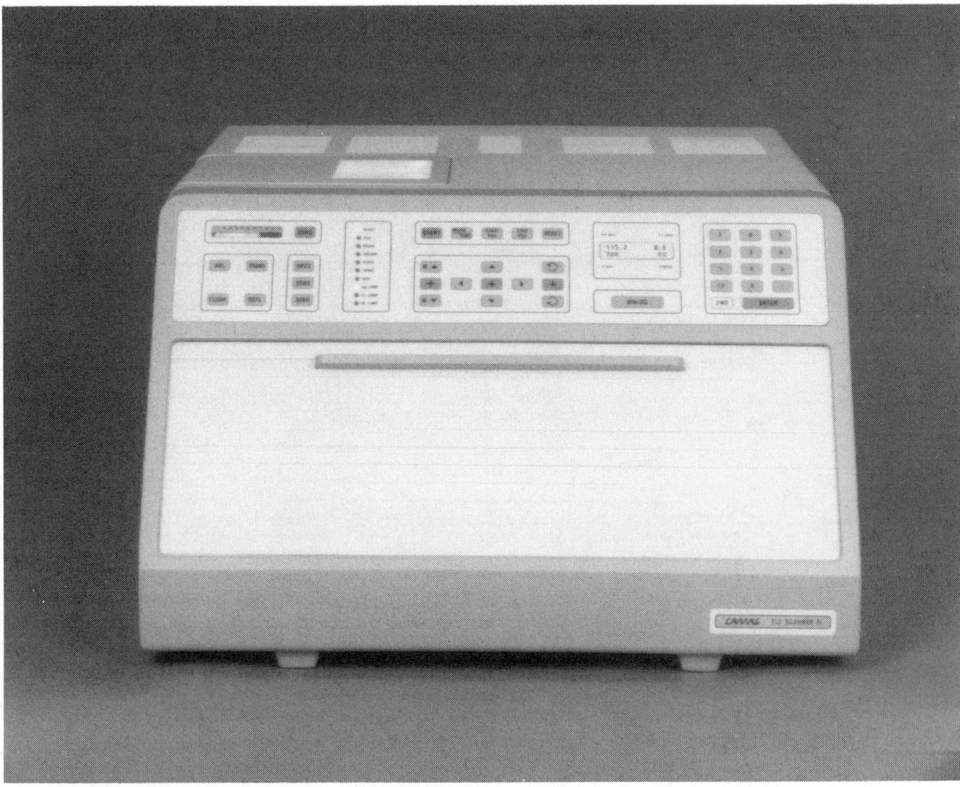

Figure 6 CAMAG TLC scanner II.

data must be available. A CAMAG TLC scanner II (Figure 6), additionally equipped with a power supply for the deuterium lamp, has been used. The analog signals are sent to an integrator, HP 3390 A (Hewlett-Packard), and the corresponding peak areas to an interface (Biochemie GmbH) which controls the scanner. The printer used for dialog and printout of reports is an LA 12 D, the computer a PDP 11/44 (both Digital Equipment Corp.). In addition, a balance (Sartorius 1219 MP) is used for weighing the broths. The weights are stored on the computer disk and further introduced into the results of the analyses (see Figure 7).

In general, the scanning speed is 4 mm/s. The plates are scanned perpendicular to the direction of chromatography. This mode of scanning offers the advantage of short analysis time, be-

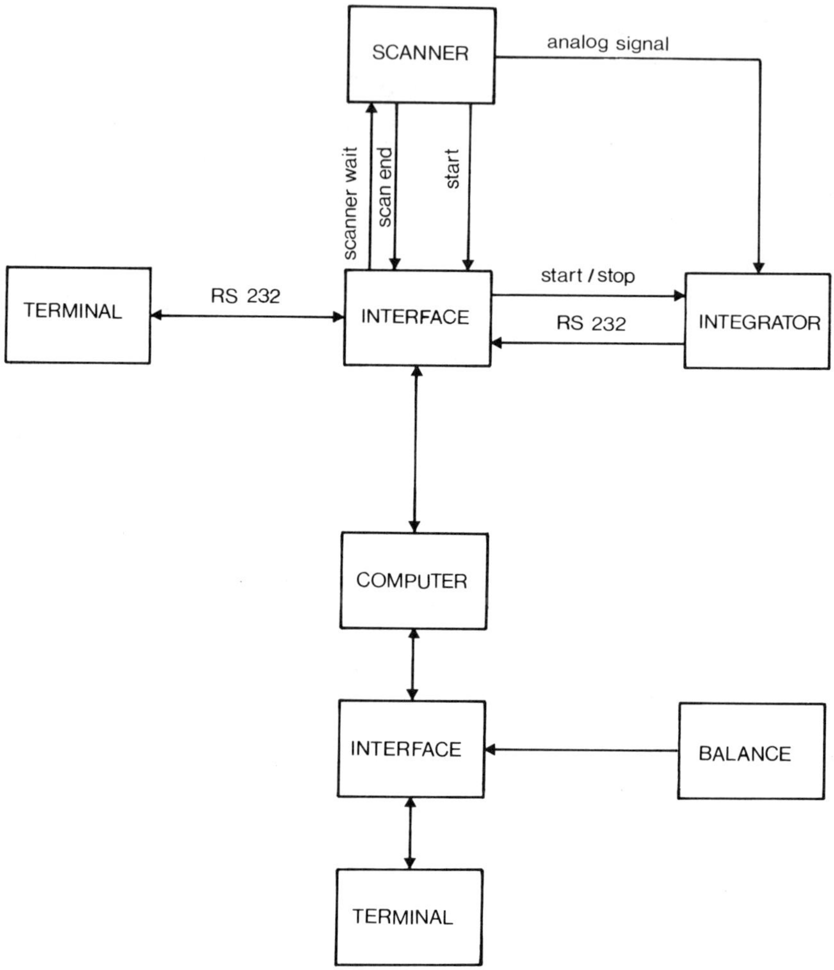

Figure 7 Scheme of measurement and evaluation, including weighing.

cause only a length of 18 cm must be scanned (measuring in the direction of chromatography would take about 120 cm). If desired, the plate can be measured automatically in the direction of chromatography.

The question of the optimal calibration function for quantitative HPTLC is, and will continue to be, a disputed issue as long as this analytical technique is performed. The usual functions are peak area against concentration, square of the peak area against concentration, logarithm of the peak area against logarithm of the concentration, and many more. In order to create space for as many samples as possible, the number of calibration standards should be kept low. Therefore we use three standard concentrations for construction of the calibration line. The fitting is done by the so-called percentage method, which is advantageous in cases where there is a wide concentration range (9). The least-squares method used before equalizes the difference between the experimental *y* values and their corrected counterparts. The constancy of the errors over the entire range is in contrast to an increase of the relative error, especially in the lower concentration range. To minimize such errors the squares of the relative errors of the *x* values are used instead of the squares of the *y* values (percentage method) (10).

$$y = a + bx$$

$$x = \frac{y - a}{b}$$

$$\frac{y_i - a}{b} - x_i = f_i \text{ (absolute)}$$

$$\frac{y_i - a}{bx_i} - 1 = \frac{f_i}{x_i} = s_i \text{ (relative)}$$

For Σs_i^2 = Min! it follows that

$$b = \frac{\sum 1/x_i^2 \sum y_i^2/x_i^2 - \left(\sum y_i/x_i^2\right)^2}{\sum 1/x_i^2 \sum y_i/x_i - \sum 1/x_i \sum y_i/x_i^2}$$

$$a = \frac{\sum y_i/x_i^2 - b\sum 1/x_i}{\sum 1/x_i^2}$$

Table 1 shows clearly that the percentage method enables more precise analyses than the least-squares method, which is statistically verified at the 99.9% level (T = 3.88).

Table 1 Accuracy of Calibration Lines

g up to about 2.5%		g more than about 2.5%	
Least-squares method	Percentage method	Least-squares method	Percentage method
1.6	1.5	2.7	2.1
2.7	1.6	6.0	3.1
1.5	1.5	3.1	3.1
1.8	1.8	4.2	2.1
1.7	1.7	2.9	2.9
1.6	1.1	3.4	2.2
1.2	1.3	3.0	2.9
1.9	1.0	4.9	3.1
2.6	1.8	2.3	2.3
1.4	1.4	2.4	1.8
1.8	1.7	2.2	2.2
2.1	1.4	4.1	3.3
1.4	0.9	3.8	2.4
1.9	1.9	3.9	2.5
1.8	1.5	2.4	2.5
1.8	1.5	3.2	2.6
1.8	1.7	2.4	2.5
1.8	1.0	3.2	2.4
2.0	2.0	4.0	2.9
1.9	1.4	2.9	2.7
Σ = 36.3	Σ = 29.7	Σ = 67.0	Σ = 51.6
Δ = 6.6 = 22.2%		Δ = 15.4 = 29.8%	
T = 4.00		T = 4.13	

III. SAMPLE PREPARATION

The first step is to obtain a homogeneous, representative sample. The fermentation broth can be of low or high viscosity, and therefore the analyst must cope with different sedimentation rates of insoluble particles, foaming, etc. Preparation methods depend on whether the whole broth or the filtrate is used for analysis. According to the substance to be determined, and the biological matrix of the fermentation liquid, a cleanup of the sample is necessary. Methods for this are extraction (of matrix or substance) and precipitation (usually of matrix). The preparation should be a fast one-step method ensuring high stability of the substance to be analyzed. On the other hand, the possibility should be considered of combining chromatography with cleanup (for example, precipitation of proteins on the place of application) and, finally, of choosing a selective derivatization reagent. The aim of this strategy should be to have spots on the HPTLC plate sufficiently separated from neighboring zones or tracks of the matrix, so that scanning can be done perpendicular to the direction of chromatography.

IV. PROCESS CONTROL OF A PENICILLIN V FERMENTATION WITH QUANTITATIVE HIGH-PERFORMANCE TLC

The following substances can be analysed with quantitative HPTLC:

Final product	By-product	Other components of fermentation broth
Penicillin V	4-Hydroxy-Penicillin V	Phenoxyacetic acid Sucrose Lactose Soya oil

The experimental conditions for the determination of the above components are described as follows.

A. Determination of Penicillin V, 4-Hydroxy-Penicillin V, and Phenoxyacetic Acid (5)

The fermentation broth is diluted with acetonitrile and filtered to remove suspended solid particles. The standard solutions contain-

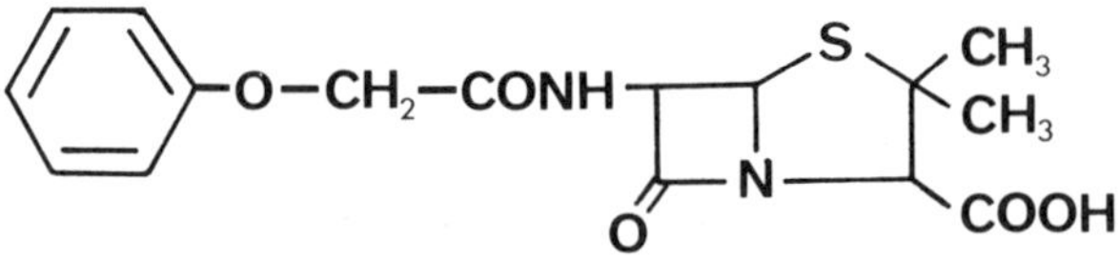

Figure 8 Structure of penicillin V.

ing known concentrations of penicillin V (for structures see Figure 8), 4-hydroxy-penicillin V, and phenoxyacetic acid are prepared in the same solvent mixture as the samples.

In order to prevent opening of the β-lactam ring, treatment of the sorbent at the origin with acetic acid (16) is recommended before application of the samples. The chromatography is carried out on silica gel 60 F254 HPTLC plates (No. 5642 from E. Merck) in a solvent mixture consisting of toluene-ethyl acetate-acetic acid (40:40:20, v/v). The developed plate is dried briefly at 120°C and evaluated in a scanner at 268 nm in the direction of the chromatography (Figure 9). The R_f values for 4-hydroxy-penicillin V, penicillin V, and phenoxyacetic acid were 0.34, 0.44, and 0.55, respectively. The accuracy of the assay depends on the amount applied on the plate. The coefficient of variation (CV) may range from 1.5 to 3,0%, but is higher (6.6% for 4-hydroxy-penicillin V) due to the low concentration.

B. Determination of Lactose and Sucrose (11)

The fermentation broth is diluted with methanol and filtered. The standard solutions are prepared with the same solvent. The chromatography is performed on silica gel 60 F254 HPTLC plates (No. 5642 from E. Merck) in a solvent mixture consisting of 1-butanol-acetic acid-water (40:50:7.5, v/v). The developed plate is heated to 120°C for 30 min. The plate is then sprayed with a reagent consisting of aniline, diphenylamine, phosphoric acid, and methanol (2 ml, 2 g, 10 ml, and 88 ml, respectively). Lactose appears at R_f 0.16 as gray-blue spots and sucrose at R_f 0.26 as brown-green zones when heated at 110°C for 15 min. Both components can be evaluated quantitatively on a TLC scanner at 523 nm (Figure 10). The accuracy was determined at a CV ranging from 1.5 to 2.0%.

C. Determination of Soya Oil (11)

The fermentation broth samples are extracted with chloroform. The chloroform phase is diluted if necessary. Standard solutions

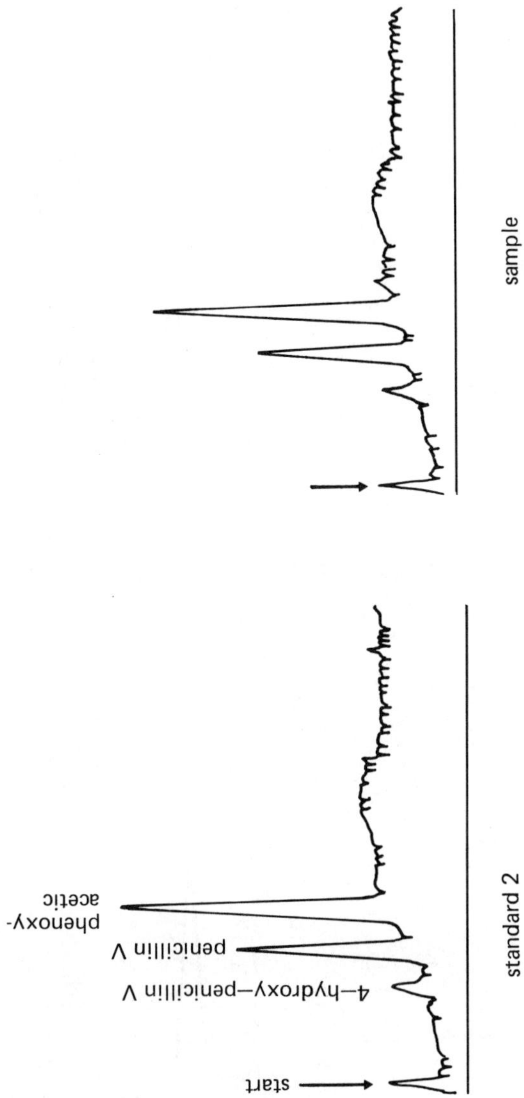

Figure 9 Densitometric scan of a broth sample along the corresponding standard derivatives.

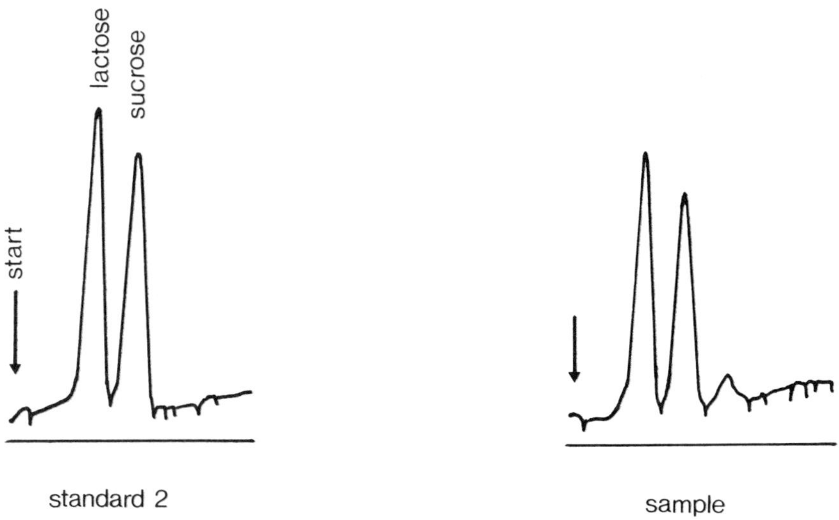

Figure 10 Densitometric scan of lactose and sucrose in a broth sample and the corresponding standard.

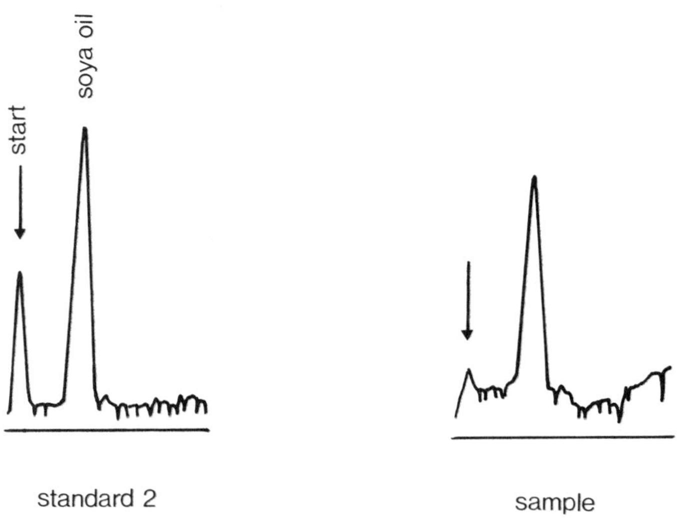

Figure 11 Densitometric scan of soya oil in a fermentation broth and the corresponding standard.

are prepared in the same solvent. The same plates are used as in the previous two methods. The chromatographic solvent is a mixture of petroleum ether-diethyl ether-acetic acid (90:10:1, v/v). The developed plate is dried for 10 min at 120°C, then sprayed with the reagent solution (3 g ammonium molybdate + 25 ml water + 30 ml 0.1 N HCl + 15 ml 60% perchloric acid + 70 ml methanol). Blue spots develop at 115°C in 3 min at R_f 0.16. The quantitative evaluation is performed on a TLC scanner at 630 nm (Figure 11). The accuracy was reported at a CV of about 3%.

V. ANALYSIS OF DIFFERENT SUBSTANCES IN FERMENTATION MEDIA WITH QUANTITATIVE TLC AND HIGH-PERFORMANCE TLC

A. Amino Acids

Provided the amino acids to be analyzed can be separated sufficiently from each other or, the amino acid(s) of interest is (are) present in far higher concentration than the other(s), HPTLC methods can easily be performed. Today, for the derivatization of the separated amino acids fluorogenic reagents such as Fluram are used. As an example the determination of methionine is described (4).

The fermentation liquid is diluted with aqueous methanol (50%) and centrifuged. A volume of 250 μl of the standards (0.4, 1.2, and 2.0 mg methionine per milliliter of 50% methanol) and samples are spotted with Pt-Ir capillaries onto the silica gel HPTLC plates (E. Merck). The separation is performed with the solvent 1-butanol-acetic acid-water (80:20:20) over a distance of 4 cm, requiring about 30 min.

The plate is dried (30 min at 120°C) and sprayed with borate buffer (pH 8.0), heated (10 min at 120°C), sprayed with Fluram solution (45 mg in 150 ml acetone), and dried (3 min at 120°C) again. The R_f value is 0.41. After 20–30 min the spots are measured with a CAMAG TLC scanner II. The excitation wavelength is 365 nm. The fluorescence is measured in the direction of chromatography—after inserting a cutoff filter for 400 nm—at a speed of 2 mm/s (see Figure 5).

B. Citric Acid

Filtrates of fermentation broth have been analyzed (12) without further pretreatment. The standards are aqueous solutions (1, 2, 3, 4, and 5 mg/ml) of trisodium citrate. Standards and samples

(1 μl) are spotted onto silica gel plates (20 x 20 cm, E. Merck), air-dried for 30 min, and developed in 1-butanol-acetone-ammonia-water (35:25:20:10). The distance of migration is 14 cm. The plates are air-dried, sprayed with a solution of 4 g 4-dimethyl-aminobenzaldehyde in 100 ml acetic anhydride, and heated at 150°C for 3 min.

The red spots are quantitated with CAMAG scanner at 565 nm. There is a linear relationship between the peak areas and the concentrations from 1 to 5 mg/ml. The coefficient of variation is between 2 and 4%; the recovery of added citric acid lies between 94 and 104%.

C. Dicarboxylic Acids

Usually—for qualitative purposes—separated dicarboxylic acids are visualized by spraying with a pH indicator. These methods result in moderate sensitivity, due to the colored background of the plate. Acridine and its compounds show a pH-dependent fluorescence spectrum, giving distinct differences in color between the spot and the background of the plate (13).

Separation has been done on silica gel plates (20 × 20 cm, thickness 0.5 mm, self made). The solvent system is chloroform-methanol-ammonia (30:20:10). One microliter each of samples and standards (0.1% in acetone) are spotted and developed (10 cm, 75 min). The plate is dried for 30 min at 120°C and sprayed with bis-3,6-dimethylaminoacridine (0.05% in ethanol). Excitation at 254 or 366 nm results in red spots and a yellow-green background. No specific measuring wavelength is recommended. The R_f values are as follows:

Oxalic acid	0.12
Malonic acid	0.39
Succinic acid	0.56
Glutaric acid	0.67
Adipic acid	0.74
Sebacic acid	0.88

D. Polyalcohols

Glycerol and sorbitol can be analyzed by means of quantitative HPTLC (14) on silica gel HPTLC layers (E. Merck). The spotting volume for samples and standards is 1 μl. The concentration is about 500 μg/ml. As solvent system a mixture of ethyl acetate, pyridine, water, acetic acid, and propionic acid (50:50:10:5:5) is recommended.

The migration distance is 7 cm. The plate is dried and derivatized by dipping in a mixture of 5 ml saturated lead tetraacetate in acetic acid, 5 ml 1% 2,7-dichlorofluorescein in ethanol, and 190 ml toluene, and heating for 3 min at 110°C. The R_f values are 0.41 for sorbitol and 0.73 for glycerol. The wavelength for excitation of fluorescence is 313 nm. A cutoff filter FL 390 is used for selecting the emission wavelengths.

VI. ANALYSIS OF ANTIBIOTICS BY QUANTITATIVE TLC AND HPTLC

In this chapter the determination of some antibiotics by means of quantitative TLC and HPTLC is described. Of course, no attempt is made to give a complete list of antibiotics. Usually, the TLC methods presented can easily be transformed into HPTLC methods. The origin of the samples to be analyzed is widespread: fermentation broths, ointments, urine, and pure products.

A. Ampicillin

Ampicillin (Figure 12) has been determined in urine samples (15). To 10 ml of urine, 3 g of ammonium sulfate was added. This solution was extracted three times with 10 ml chloroform containing 166 mg benzalkonium chloride. The extracts were evaporated in order to achieve a volume of less than 10 ml. After transfer to a 10-ml volumetric flask, the flask was filled to volume with chloroform.

The separation was performed on HPTLC plates of silica gel 60 F254, 10 × 10 cm, in a solvent system consisting of dioxane-water-1-butanol-formic acid (70:15:15:1.25). The standard was ampicillin trihydrate dissolved in dioxane-water (4:1).

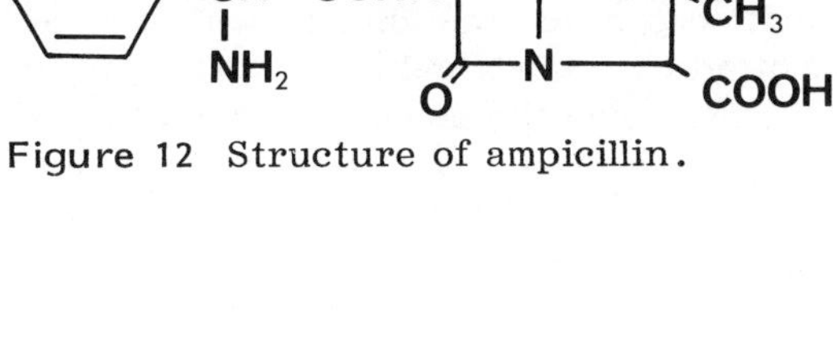

Figure 12 Structure of ampicillin.

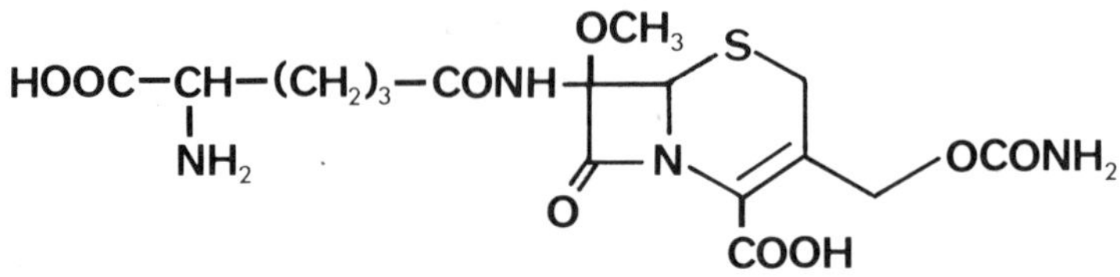

Figure 13 Structure of cephamycin C.

The spotting volume was 1 μl. After a migration distance of 7 cm, the plates were air-dried, sprayed with ninhydrin reagent, and heated for 5 min at 110°C. The R_f value of ampicillin was found to be 0.65.

The evaluation was performed at 480 nm on a Zeiss TLC scanner connected to a Varian A-25 recorder and an Apple II microcomputer. A linear range of detection was given from 0.05 to 1.00 μg per spot. The coefficient of variation did not exceed 3%. The limit of quantitation was 0.05 mg/ml urine.

B. Cephamycin C

The fermentation broth was centrifuged, and the supernatant was used for the analysis (16). On the spotting points on 20 × 20 cm silica gel 60 F254 plates (E. Merck) 10 μl of acetic acid was applied in order to prevent decomposition of the β-lactam ring (for structure, see Figure 13), followed by 5 μl of the samples and standards after drying the plate.

Eight samples and two standards (0.02 and 0.6 g/l) were applied on one plate. The separation was performed with the solvent ethanol-acetic acid-ammonia (6:3:1, v/v). After drying the plate for about 1 h, scanning was done at 273 nm. The linearity of the calibration line was given from 0.2 to 20 μg per spot. An accuracy of 1.6 to 3.1%, and a recovery of 97.6% were reported. The correlation with the HPLC method was $r^2 = 0.99124$.

C. Daunorubicin

Daunorubicin (Figure 14) is a fermentation product of *Streptomyces penceticus*. In order to cleave the glycoside from daunorubicin (17), 10 ml of broth was treated for 45 min at 50°C with 300 mg of oxalic acid. One milliliter of a clarified solution was diluted with 1 ml of acetone, and 2–5 μl was spotted on a silica gel plate. The concentration of standard was 20 μg/ml. For the separation a mixture of chloroform-methanol-acetic acid (65:10:10, v/v) was used. After drying for 1 min with a hair dryer the plate was measured at 480 nm

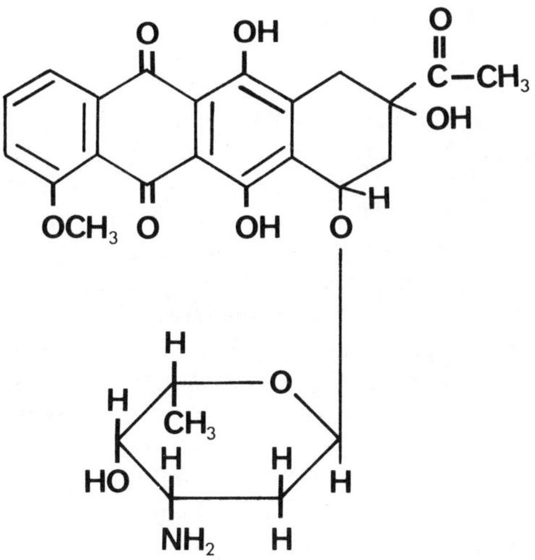

Figure 14 Structure of daunorubicin.

(R_f 0.37). Accurate results were obtained between 10 and 100 μg/ml. The recovery was more than 90%.

D. Gentamicin

Gentamicin, a complex mixture of the components C_{1a}, C_1, and C_2 (see Figure 15), is a fermentation product of *Micromonospora purpurea*. The compounds differ from each other in the degree of methylation at the 6'-position. The occurrence of component C_{2a} is not yet mentioned in papers.

The three amino groups of the molecule react with 4-chloro-7-nitrobenzo-2-oxa-1,3-diazole (NBD-C2), giving a product which

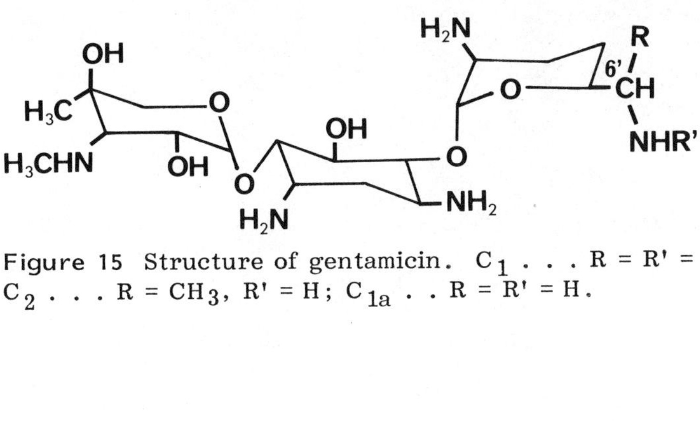

Figure 15 Structure of gentamicin. C_1 . . . $R = R' = CH_3$; C_2 . . . $R = CH_3$, $R' = H$; C_{1a} . . $R = R' = H$.

can be excited at 420 nm. The fluorescence is measured at 530 nm (18). The fermentation broth is acidified to pH 2.0 with H_2SO_4, centrifuged, and the supernatant adjusted to pH 12 with NaOH. Samples and standards 5 μl each are applied (data pair technique) on silica gel plates and chromatographed in the lower layer of methanol-chloroform-ammonia (1:1:1, v/v), air-dried, and dipped into a methanolic NBD-Cl solution (0.24 mg/ml) for 2 s. After heating at 120°C for 10 min, the plate is cooled and rechromatographed in methanol in order to remove excess reagent. A linear relationship between peak area and concentration is given from 2 to 500 ng per spot.

Another method of derivatization of the gentamicin components by means of ninhydrin has also been described (19). The separation is performed on silica gel layers (aluminum sheets) with the solvent system chloroform-methanol-ammonia (1:1:1, v/v) (lower layer) over a distance of 15 cm. Standards C_1, C_{1a}, and C_2 and samples (10 mg/ml) are applied in stripes 1.5 cm long. After separation (distance 15 cm), the sheet is dried for 30 min at 120°C and sprayed with ninhydrin (1 g ninhydrin + 0.25 g cadmium acetate + 10 ml acetic acid + 50 ml ethanol), and heated for 2 h at 120°C. The spots are measured with an ERI-65 scanner at 510 nm. The relative standard deviations range from 4.1 to 5.3%.

Finally, a derivatization of the amino groups with Fluram was published (20). The layer was silica gel 60 (20 × 20 cm, E. Merck), the solvent system chloroform-methanol-ammonia (1:1:1, v/v) (lower layer, dried with 1.5 g K_2CO_3 for 20 min). The separation distance was 10 cm (1 h).

Gentamicin sulfate (66.7 mg), dissolved in 10 ml borate buffer (pH 8.0), is stirred with 10 ml of chloroform. One microliter of the upper layer is spotted onto the plate. The sample is treated in the same manner. After the separation the plate is dried in air for 1/2 h and sprayed with a borate buffer, then several times with a Fluram solution (0.04% in anhydrous acetone). Finally, for stabilization of the fluorescence, the plate is treated with a mixture of triethylamine and methylene chloride (10:90, v/v). Three hours later excitation is given at 365 nm. The emission is measured at 460 nm (Schoeffel SD 3000). The R_f values are: C_{1a}, 0.32; C_2, 0.40; C_1, 0.45.

E. Gramicidin

Tyrothricin, formed by *Bacillus brevis*, is a mixture of about 20% gramicidin, a linear polypeptide, and about 80% tyrocidine, a cyclic polypeptide. Gramicidin and tyrocidine themselves are not homogenous substances, but mixtures of structurally closely related compounds which do not need any further differentiation.

The fermentation broth is adjusted to pH 4.0 by addition of HCl, followed by centrifugation, extraction of the sediment with ethanol, and a further centrifugation (2). Samples and standards (400, 1100, and 1800 μg gramicidin/ml ethanol) are spotted (250 nl) on a silica gel HPTLC plate (E. Merck). The solvent system is acetic acid-butyl acetate-1-butanol-methanol-water (20:40:7.5:2.5:12, v/v). The migration distance is 3 cm. The plate is dried for 30 min at 115°C, sprayed with a reagent (2 g of 4-dimethylaminobenzaldehyde + 25 ml HCl + 75 ml ethanol), heated for 5 min (115°C), and scanned at 570 nm. The spots are violet (R_f 0.78). The calibration line is peak area squared vs. concentration. The coefficient of variation is between 2 and 4% (N = 10).

F. Griseofulvin

This antibiotic (Figure 16) is a fermentation product of *Penicillium urticae*. It is extracted from fermentation broth with ethyl acetate (21). The samples and standards (0.2, 0.6, and 1.0 mg/ml) are spotted (250 nl) on a silica gel HPTLC plate (E. Merck), and chromatographed in butyl acetate-acetone (4:1, v/v). The migration distance of 3 cm is reached after 5 min (R_f 0.5). The plate is dried for 30 min at 120°C. Griseofulvin is excited to fluorescence at 295 nm. A cutoff filter allows scanning from 390 nm upward. The squares of the peak areas and the concentrations are used for the calibration line. The coefficient of variation is 1 to 2% (N = 10).

In the presence of by-products, such as dechlorogriseofulvin, a separation is performed on reversed phase layers (Nano SIL C_{18}-50, Macherey-Nagel) with the solvent methanol-water (4:6, v/v) over a distance of 6 cm.

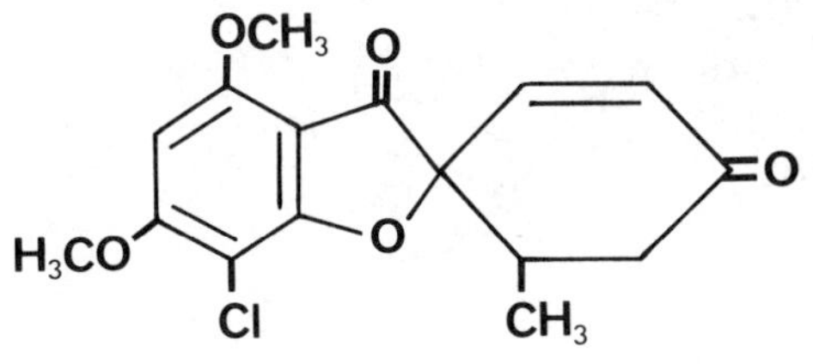

Figure 16 Structure of griseofulvin.

G. Neomycin, Bacitracin, Polymyxin

These three polypeptide antibiotics can be separated with the following solvent system on silica gel layers (20 × 20 cm): 1-butanol-acetic acid-water (40:10:20, v/v) over a separation distance of 11 cm (4.5 h) (20). As a standard solution, 3.6 mg polymyxin, 17.6 mg bacitracin, and 10.0 mg neomycin sulfate are dissolved in 10 ml water. The sample, an ointment, is extracted with 5 ml chloroform plus 5 ml water. The aqueous phase is used for analysis. Standard and sample (1 μl each) are spotted. After the separation, the plate is dried for 4 h in the air and sprayed with borate buffer (pH 8.0) and Fluram (0.08% in anhydrous acetone).

All these derivatives are excited at 365 nm. The emission is scanned at 460 nm. The R_f values are: bacitracin A, 0.34; bacitracin B, 0.32; polymyxin, 0.32; neomycin sulfate, 0.0.

The by-product bacitracin B is present in negligible traces. Neomycin can be detected at the starting point due to the strong specificity of the derivatization.

H. Spiramycin

Spiramycins I, II, and III are 16-member ring macrolide antibiotics (Figure 17), obtained by fermentation of *Streptomyces ambofaciens*. An HPTLC-method has been developed (22) for bulk powders, tablets, and suppositories. Silica gel HPTLC plates (E. Merck) have been used. The eluent is the lower layer of the mixture methylene chloride-1-propanol-ammonia (93.5:5:1.5, v/v). The separation is done at 30°C (over 6 cm).

The standard solutions are: spiramycin I (0.5–1.5 mg/ml), spiramycin II (0.25–0.75 mg/ml), spiramycin III (0.22–0.66 mg/ml),

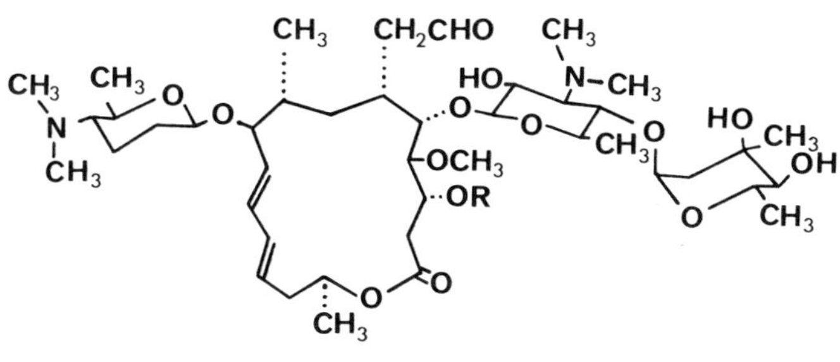

Figure 17 Structures of spiramycins. Spiramycin I, R = H; II, R = $COCH_3$; III, R = $COCH_2CH_3$.

Table 2 Tetracycline Compounds

Compound	Structure	R_1	R_2	R_3	R_4	R_5	R_6
Tetracycline (TC)	I	H	CH_3	OH	H	$N(CH_3)_2$	H
Oxytetracycline (OTC)	I	H	CH_3	OH	OH	$N(CH_3)_2$	H
Chlorotetracycline (CTC)	I	Cl	CH_3	OH	H	$N(CH_3)_2$	H
Doxycycline (DC)	I	H	CH_3	H	OH	$N(CH_3)_2$	H
4-Epitetracycline (ETC)	I	H	CH_3	OH	H	H	$N(CH_3)_2$
Anhydrotetracycline (ATC)	II	H	—	—	—	$N(CH_3)_2$	H
4-Epianhydrotetracycline (EATC)	II	H	—	—	—	H	$N(CH_3)_2$

and as internal standard trifluperidol (3 mg/ml) dissolved in benzene and methanol (9:1, v/v). Sample volumes of 100 or 200 nl have been spotted. The scanning is done at 232 nm. For calibration, peak heights versus concentration have been plotted. The coefficient of variation did not exceed 5% ($N = 5$). The R_f values are 0.21, 0.35, 0.41, and 0.61 for the internal standard.

I. Tetracyclines

A summary of tetracycline compounds appears in Table 2. For the separation of the tetracyclines (Figure 18) silica gel plates—which are predeveloped with a saturated aqueous solution of Na_2EDTA, dried at room temperature for 1 h, and activated at 130°C for 2 h—are used (23). Samples and standards, (100 nl of each) are spotted on the plate. The concentration is about 10 mg/ml methanol.

The solvent systems are: (a) chloroform-methanol-5% Na_2EDTA aqueous solution (65:20:5, v/v), lower layer; (b) propanol-2-ethyl acetate-5% Na_2EDTA aqueous solution (3:4:7, v/v) upper layer; and (c) acetone-5% Na_2EDTA aqueous solution (10:1).

After the separation the plate is scanned in the dual-wavelength mode.

λ_{sample} = 260 nm for TC, OTC, CTC, DC, ETC $\lambda_{reference}$ = 600 nm

λ_{sample} = 450 nm for ATC, EATC $\lambda_{reference}$ = 650 nm

with scanning beam 0.25 × 9.0 mm.

The R_f values of the tetracyclines are shown in Table 3. According to the tetracyclines to be separated, a proper solvent system must be selected. A linear relationship between peak area

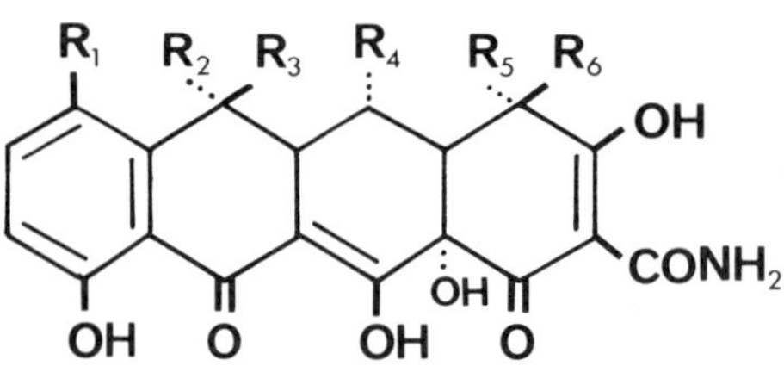

I

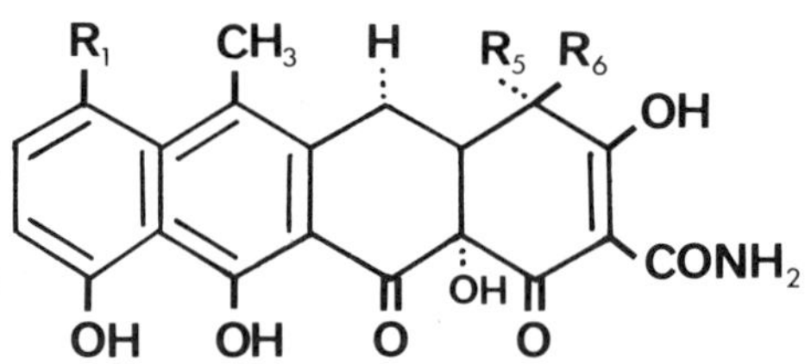

II

Figure 18 Structures of tetracyclines.

Table 3 R_f Values of the Tetracyclines

	Solvent		
Compound	A	B	C
ATC	0.87	0.83	0.94
EATC	0.46	0.22	0.38
CTC	0.44	0.28	0.56
TC	0.38	0.18	0.45
DC	0.26	0.28	0.45
OTC	0.12	0.35	0.60
ETC	0.10	0.11	0.38

and concentration is given between 0.2 and 1.0 μg/spot. Poor resolution can be overcome by changing the solvent and the wavelength of measurement.

Another paper (24) describes the densitometric determination on kieselguhr layers impregnated with Na_2EDTA (self-prepared plates). The solvent system is acetone-ethylacetone-water (20:10:3, v/v). Scanning is performed at 400 nm for TC and 363 nm for OCT and CTC. The standard is dissolved in methanol (0.25%). Two microliters are spotted. The R_f values are: TC, 0.70; OCT, 0.65; and CTC, 0.60. The linearity of the calibration line is given between 0.4 and 2.0 μg/spot. The coefficient of variation ranges from 4 to 7%.

J. Turimycin

The turimycin complex (H_2, H_3, H_4, H_5) (Figure 19) is a metabolic product of *Streptomyces hygroscopicus*. The standards and samples contain about 2 mg/ml methanol each. RP 18 plates (E. Merck) are used (22). The solvent system is acetonitrile-methanol-ammonium acetate (5% aqueous) (40:30:30, v/v), and the temperature for the sepration is 35°C. The migration distance is 5 cm, the volume of spotting 100 or 200 nl. Scanning is performed at 232 nm. The R_f values were found as follows: H_5, 0.22; H_4, 0.31; H_3, 0.42; and H_2, 0.58.

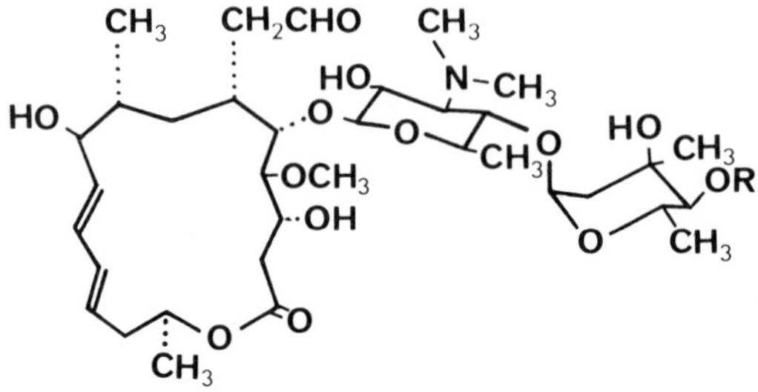

Figure 19 Structures of turimycins. H_2, R = $COCH_3$; H_3 R = $CHCH_2CH_3$; H_4, R = $COCH_2CH_2CH_3$; H_5, R = $COCH_2CH(CH_3)_2$.

VII. FUTURE ASPECTS

Usually, the analyst in industry must minimize analysis time and costs as well as improve accuracy. Therefore it is necessary to evaluate the potential of TLC and especially HPTLC in comparison to competitive methods such as HPLC. What has changed in the past three years? On the one hand, the instrumentation of HPTLC has been improved and expanded, and on the other, super-speed LC has been introduced. That means a step forward for both chromatographic methods.

Nevertheless, what should still be improved regarding quantitative HPTLC? The analyst's answer is very simple: the speed. This includes the speed of spotting, separation, and scanning. Of course, manual spotting has been replaced by instrumental spotting, but the instruments available in the future must be faster by a factor of 4.

The technique of thin-film chromatography (25), a high-speed separation introduced by E. Merck at Bad Dürkheim in 1975 (26), and the paper of Andreev (27) could be the catalysts for creating "fast" layers.

The speed of scanning must be increased (up to 1 m/min), which means that scanners and integrators must cope with at least three peaks per second. The acquisition and evaluation of data will be a problem easy to solve.

If these objectives can be fulfilled in the near future, quantitative HPTLC will be a still more valuable chromatographic method.

ACKNOWLEDGMENTS

I am indebted to Miss Veronika Bichler, Miss Anita Moser, and Miss Edith Trainer for their excellent technical assistance, to Mr. Helmut

Wolff for writing the computer programs, and finally to Mr. Dieter Ascher and Mr. Josef Lettenbichler for the electronic equipment.

REFERENCES

1. F. Kreuzig. Z. Anal. Chem. *281*, 135 (1976).
2. F. Kreuzig. Z. Anal. Chem. *282*, 447 (1976).
3. H. Bethke, W. Santi, and R. W. Frei. J. Chromatogr. Sci. *12*, 392 (1974).
4. F. Kreuzig. GIT Suppl. Chromatogr. *82*, 46 (1982).
5. F. Kreuzig. Oesterr. Chem. Z. *4*, 96 (1982).
6. S. Ebel, E. Geitz, J. Hocke, and M. Kaal. Kontakte (Merck) *3*, 19 (1981).
7. F. Kreuzig. Z. Anal. Chem. *282*, 457 (1976).
8. F. Kreuzig. Chromatographia. *13*, 238 (1980).
9. J. Knecht and G. Stork. Z. Anal. Chem. *270*, 97 (1974).
10. F. Kreuzig. In: *Instrumental HPTLC*, W. Bertsch, S. Hara, R. E. Kaiser, and A. Zlatkis (eds.), p. 224, Hüthig, Heidelberg, 1980.
11. F. Kreuzig. J. Liq. Chromatogr. *6*, 1227 (1983).
12. H. Socic and V. Gaberc-Porekar. Z. Anal. Chem. *309*, 114 (1981).
13. C. Sarbu, M. Horn, and C. Marutoiu. J. Chromatogr. *281*, 345 (1983).
14. R. Klaus and R. Ripphahn. J. Chromatogr. *244*, 99 (1982).
15. A. Mrhar, F. Kozjek, M. Prosek, and A. Dobovisek. J. Chromatogr. *277*, 251 (1983).
16. L. R. Treiber. J. Chromatogr. *213*, 129 (1981).
17. H. J. Isaaq, N. H. Risser, and A. Aszalos. J. Liq. Chromatogr. *2*, 533 (1979).
18. P. Kabasakalian, S. Kalliney, and A. W. Magatti. Anal. Chem. *49*, 953 (1977).
19. I. Török and T. L. Paal. J. Chromatogr. *246*, 356 (1982).

20. M. Breitinger, H. Paulus, and W. Wiegrebe. Dtsch. Apoth. Ztg. *36*, 1699 (1980).

21. F. Kreuzig. Oesterr. Chem. Z. *4*, 108 (1980).

22. G. A. Gens, W. Van den Bossche, and P. De Moerloose. J. High-Resolution Chromatography & Chromatography Communications *3*, 433 (1980).

23. H. Oka, K. Uno, K. O. Harada, Y. Kaneyama, and M. Suzuki. J. Chromatogr. *280*, 457 (1983).

24. C. G. Gatsonis and C. A. Ageloudis. Pharmazie *37*, 649 (1982).

25. E. Cremer, F. Deutscher, P. Fill, and H. Nau. J. Chromatogr. *48*, 132 (1970).

26. R. E. Kaiser. In: *Einführung in die Hochleistungs-Dünnschichtchromatographie*, R. E. Kaiser (ed.), p. 4, Institut für Chromatographie, Bad Duerkheim, 1976.

27. L. V. Andreev. J. Liq. Chromatogr. 5, 1573 (1982).

10
Assay of Bioactive Materials in Plant Extracts

MIRKO PROSEK and MILICA KATIC Research Department, Lek Ljubljana, Ljubljana, Yugoslavia

MARKO PUKL *Department of Pharmacy, Edvard Kardelj University, Ljubljana, Yugoslavia*

I. INTRODUCTION

Throughout their history, humans have deliberately used plants as medicine. In fact, until the 19th century only natural products were used for the treatment of various diseases. Even today, in countries with developed chemical and pharmaceutical industries, many prescription drugs contain active principles obtained from plants, and many commonly used synthetic drugs have prototypes among natural products.

In some countries, different preparations based on crude plant extracts are still very popular. The complete composition and sometimes even the active principles of these products are not specifically known. The isolation and identification of single new active components from these preparations and from plants which exhibit biological activity is of interest for studies of structure-activity relations. The modification of bioactive natural products is one of the customary approaches in drug design. However, the introduction of traditional remedies into modern therapy requires standardization of production and use, precise determination of activity, and development of quality control methods. The evaluation of active components in plants and plant extracts is important not only for quality control of final preparations (some traditional remedies) but also for planning of economic cultivation and exploitation of some industrially important plants.

Assay of bioactive materials can be performed by means of different analtyical procedures. According to the present state of analytical chemistry, chromatography, especially thin-layer chromatography (TLC), is an indispensable tool. In this chapter we will describe our experience with the use of an on-line densitometer for the evaluation of some plant extracts.

II. ANALYTICAL PROCEDURES FOR EVALUATING PLANT EXTRACTS

A. *Papaver bracteatum*

The great scarlet poppy (*Papaver bracteatum* Lindl.) has not been used in traditional medicine, and it did not attract much attention before 1960. This herb belongs to the poppy family (*Papaveraceae*), which is known to contain alkaloids. Certain strains of *P. bracteatum* are very rich sources of thebaine, from which codeine, a widely employed analgesic and antitussive, and some narcotic antagonists can be synthesized. Earlier, codeine was almost exclusively derived from opium or from morphine. The great scarlet poppy produces mainly thebaine but, depending on the strain, also produces small amount of some other alkaloids. It does not contain opium. Morphine can be derived from thebaine only via a complex chemical procedure. It is hoped that substitution of this species for the opium poppy (*P. somniferum*) will help to reduce the international opium abuse problem (1).

The content of alkaloids in the great scarlet poppy may vary during the vegetation period from 0.2 to 1.7% in the dried capsules, and from 0.5 to 1.5% in the dried roots of some wild-grown plants (2). Obviously, it should be as high as possible for economic industrial exploitation. The yield of the isolated thebaine depends mainly on the strain of the plant, the growth conditions, and the vegetation period in which the plant is used. Several laboratories try to select and/or develop strains with a high content of alkaloids.

During the past few years in our laboratory, selection of the great scarlet poppy has been performed in order to enrich the content of active substance in the plant tissues. The amount of thebaine has been determined in dried and powdered capsules. The drug was extracted with 5% aqueous acetic acid. The pH of the extract was adjusted to 7.5 to 8.0 with concentrated ammonium hydroxide. The solution was applied directly to TLC plates.

Separation was performed on silica gel plates (silica gel 60, without indicator, E. Merck, No. 5721). The mobile phase was chloroform-methanol (9:1, v/v). Satisfactory separation was obtained in a traveling distance of 9 cm. By using half of the plate for one run, the number of samples per plate was doubled and the cost of plates was reduced. This is an important factor when several thousand samples must be examined.

The measurements were carried out with a Zeiss PMQ II scanner connected to an Apple II plus personal computer. For the evaluation of the plates software was used which allowed measurement, integration, and postrun calculation. The plates were scanned at a wavelength of 280 nm in the reflectance mode. Calibration with three standards, 1.5, 2.5, 3.5% thebaine in a sample, was used for each plate. The sample was spotted three times on different plates. On each plate 15 samples and 3 standards were applied. The separation was performed in a chromatographic chamber for 20 × 10 cm HPTLC plates.

The raw data from the entire plate were stored on a computer disk. Usually, when there was a large number of samples, the scanning was performed continuously under the control of one Apple. The evaluation of raw data was done on a second Apple. With this combination the output was much higher becasue all the work was performed by only one operator using both systems.

In the first step the values of area and area percent were calculated from the raw data. From these results the content of thebaine in the sample was calculated. The calibration line was computed by a method that was developed and tested in our factory some years ago (3). All the results stored in data bank were used in further work. Figure 1 shows the chromatogram of a sample.

A group of three people familiar with TLC and Apple II can analyze up to 200 samples in 10 h. This means 40 plates with 18 lanes each, or a total of 720 chromatograms of standards and samples (3 standards and 15 different samples on each plate) per week. Each plate is prepared three times (Figure 2). For this number of separations by high-performance liquied chromatography (HPLC), 30 h would be needed for 200 samples, but only one injec-

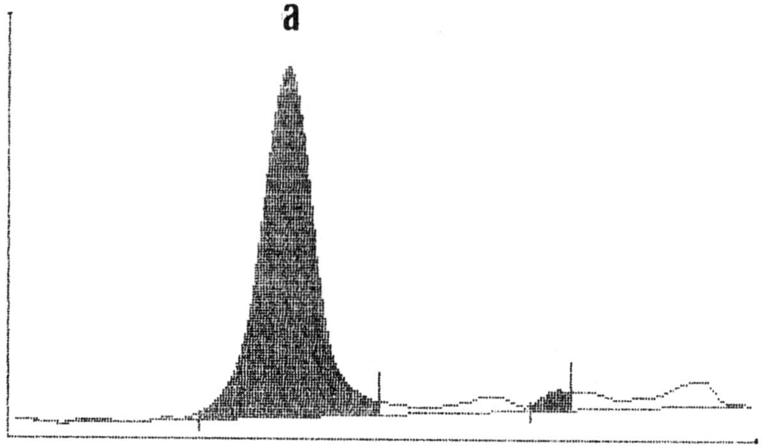

Figure 1 Densitogram of a *Papaver bracteatum* sample. (a) Thebaine. The Chromatogram was processed with poppy-pack (IFC-LEK Software). Quantitative evaluation was done in two modes: T, tangential baseline, and P, perpendicular to baseline.

tion for each sample. With three injections per sample the total time would be 100 h rather than 10 h. These numbers clearly demonstrate the utility of TLC for screening purposes.

B. Artichoke

The horticultural artichoke (*Cynara scolymus* L.), a member of the largest plant family, *Compositae* (the daisy family), has been used in folk medicine since Roman times. It is known for its cholagogic, choleretic, liver-protective, diuretic, and hypolipemic properties. The pharmacological activities of the drug are due to the presence of a number of caffeic acid derivatives. Since di- and monocaffeylquinic acids are largely hydrolyzed in the gastrointestinal tract, it is possible that the true active substance is caffeic acid. It has been demonstrated that ceffeic acid itself possesses hepatoprotective activity (4). Different caffeic acid derivatives are perhaps only suitable transport forms. All these substances are almost tasteless. The bitter taste of the leaves is due to the presence of another group of active principles, sesquiterpene lactones.

Since the artichoke is used in the food industry as a vegetable and as a source of bitter substances, different analytical methods

TLC CHROMATOGRAPHY DATA SYSTEM IFC-LEK

SELECTION: MAK

PLATE : MAK 15

DATE: 22.8

NO.	SAMPLE	AREA(P)	M	AREA(T)	M
1	S7	29102	3.56	29377	3.53
2	S5	19586	2.39	19975	2.40
3	S3	12575	1.53	12920	1.55
4	43	20424	2.49	22067	2.65
5	303	21971	2.68	22780	2.73
6	256	16670	2.03	17057	2.05
7	411	15724	1.92	16054	1.93
8	464	13518	1.65	13902	1.67
9	239	12934	1.58	13292	1.59
10	361	15555	1.90	15952	1.91
11	104	13273	1.62	13602	1.63
12	18	18693	2.28	19363	2.32
13	76	34902	4.27	35214	4.23
14	108	5136	0.62	5411	0.65
15	288	27655	3.38	28006	3.36

ANALYST: I

Figure 2 Printout for one plate. The data were send to the data bank.

for the evaluation of sesquiterpene lactones have been developed. Only gas-liquid chromatography (GLC) enables separation and quantitative determination of individual substances (5).

Sesquiterpene lactones have ultraviolet (UV) maxima at about 210 nm. These compounds represent a group of natural substances which are relatively difficult to analyze in crude plant extracts with UV detectors. Very good separation from other substances is obvious. Therefore, for HPLC and TLC special preparation of samples is advisable. In the case of TLC it is even possible to detect some sesquiterpene lactones with spray reagents (6). Colors obtained with the main sesquiterpe lactones of artichoke, cynaropicrin and grossheimin, are relatively stable.

In our laboratory we analyze the extract of the artichoke leaves. Dried and powdered samples were extracted with boiling distilled water. The extracts were filtered and evaporated. For

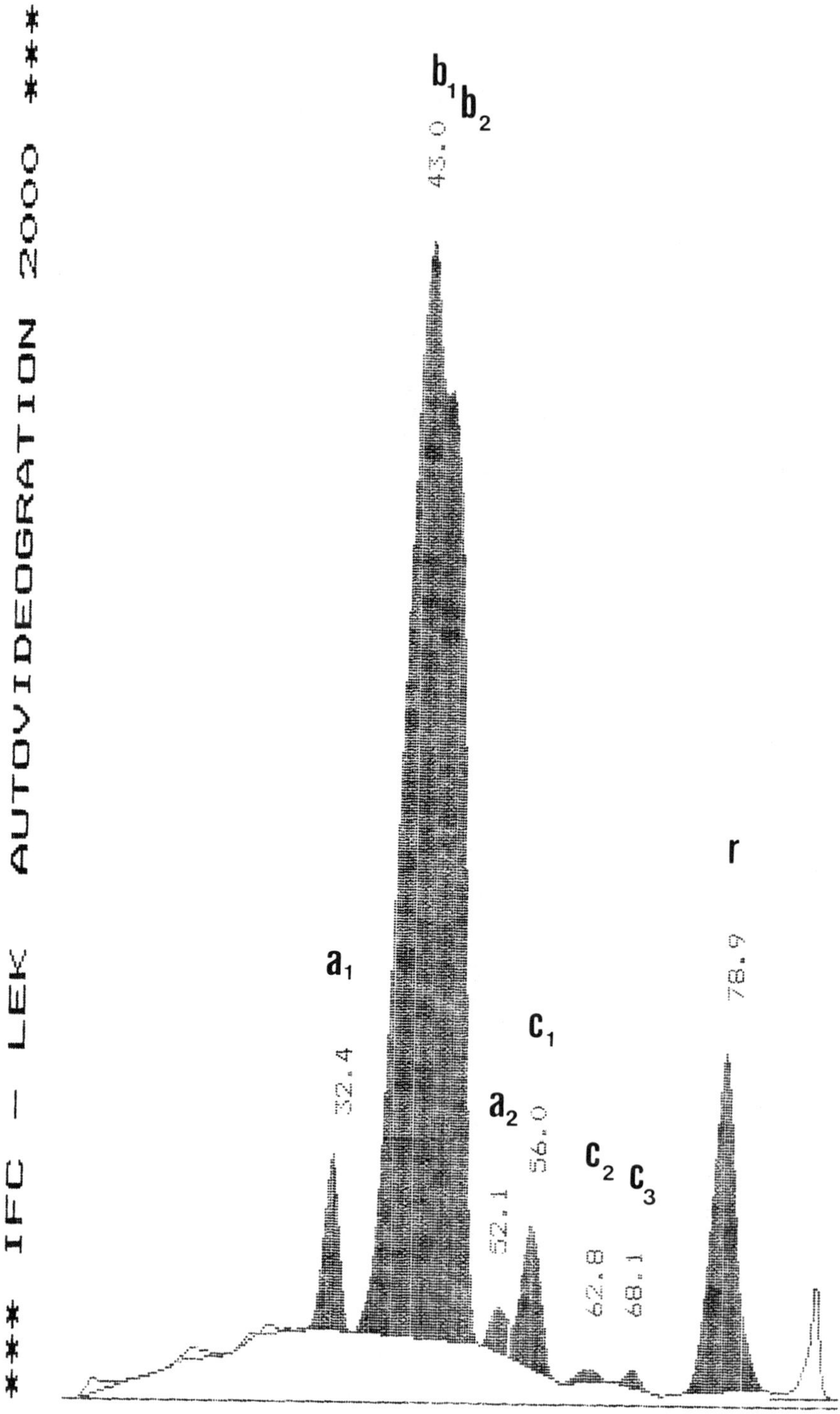
*** IFC - LEK AUTOVIDEOGRATION 2000 ***
b1b2
43.0
a1
32.4
a2
52.1
c1
56.0
c2
62.8
c3
68.1
r
78.9

quantitative TLC, the samples were dissolved in water and applied on the TLC plate.

Standards (0.1% chlorogenic acid and 0.1% cynarine in water) and samples were applied with a 1-μl micropipet. The separation was carried out on silica gel plates (E. Merck, silica gel 60, without indicator, No. 5721). The solvent system was ethyl acetate-acetic acid-formic acid-water (100:11:11:27, v/v). This system offered good separation of the phenolic compounds (7). The quantitative determination was done in two modes: (a) absorbance in the UV at 313 nm (absorption maximum of caffeic acid derivatives), or (b) after spraying with Natural Product Reagent A, in the fluorescence mode with an excitation wavelength of 313 nm and a cutoff filter of 395 nm (Figure 3). The extracts were evaluated according to the spots of chlorogenic acids (b1, b2) and cynarin (c1).

C. Silybum marianum

The Marian or milk thistle (*Silybum marianum* (L) Gaertn.) is another member of the daisy family. The seeds of *S. marianum* contain silymarin, a mixture of flavonolignans used in liver therapy. The main flavonolignan component of the silymarin, silybin, and the minor components silidianin, silicristin, and isosilybin are isomers and differ only in the linkage of the taxifolin moiety to the coniferyl alcohol. Flavanonol taxifolin alone is seldom a constituent of the silymarin mixture.

Although the Marian thistle has been known for a long time as a medicinal plant for digestive disorders, especially those of hepatic origin, it has only rarely been used in folk medicine. This plant became more interesting after 1960, when the most important substance, silybin, was isolated from silymarin. A few years later, antihepatotoxic activity of silymarin was experimentally proved, and the structures of silybin and its isomers were established (8).

The flavonolignans of *S. marianum* are among the best studied plant substances, pharmacologically and clinically. Their antihepatotoxic and hepatoprotective activities have been tested and confirmed in a large number of models of liver damage systems (9).

Figure 3 Densitogram of an artichoke extract: a(1), a(2), luteolin glucosides; b(1), chlorogenic acid; b(2), neochlorogenic acid; c(1), cinarin; c(2), c(3), other dicaffeylquinic acid derivatives; r, unknown compound under investigation (1',5'-dicaffeylquinic acid). Quantitation of the samples was done for chlorogenic acid (neochlorogenic acid) and cinarin.

A number of publications have dealt with their pharmacology, mechanism of action, and clinical applications.

One of their most impressive pharmacological activities is an antagonism against α-amanitin and phaloidin, the otherwise fatal toxins of the death cap fungus. It is interesting to note that among bioflavonoids only these substances exhibit such activity. Even the two constituents of silymarin flavonolignans, taxifolin and coniferyl alcohol, given alone, have no influence on the symptoms of poisoning with amatoxins (10). On the other hand, taxifolin, like silybin and other isomeric compounds, stimulates protein biosynthesis in liver cells in vitro (11). It is possible that this flavononol, beside flavonolignans, has an influence on regeneration of damaged liver parenchyma.

Several attempts have been made to synthesize silybin, the main component of the silymarin mixture, but *S. marianum* is still almost the only source of commercial silymarin and silybin. Therefore, it is essential for pharmaceutical companies producing preparations based on silybin or silymarin to control plant material and to select better strains containing higher amounts of these components.

Excellent separations of individual flavonolignans can be obtained by means of HPLC, and with carefully chosen mobile phases, even the diastereomeric pairs (for instance, silybin A and B and isosilybin A and B) can be separated (12). This analytical method is undoubtedly the method of choice for the analysis of final preparations, for studies of the precise composition of silymarin mixtures, and for determination of the exact ratio between some isomers. But for routine analysis of a large number of samples, as in screening for superior strains, this method is time-consuming.

Powdered samples of *S. marianum* were extracted with petroleum-ether for 3 h in a Soxhlet apparatus. The petroleum ether extract was discarded. After the samples had been dried in air, they were extracted with methanol. The methanol extracts were evaporated in vacuo.

The TLC separation was performed on silica gel plates (E. Merck, No. 5721). The mobile phase was chloroform-acetone-formic acid (9:2:1, v/v). The spots were quantified with a Zeiss chromatogram-spectrophotometer at 280 nm in the reflectance mode.

Signals were digitalized and stored on a computer disk. In one run 512 measurements were collected. After a scan of a whole plate the integration program computed the data on area and area percent. The concentrations and amounts of components were obtained with the postrun program. Three standard mixtures of different concentrations were spotted on every plate in order to create the calibration curve. Silybin, silicristin, and silidianin were used as standard components. Figure 4A shows a chromatogram of a standard mixture and Figure 4B a chromatogram of plant sample.

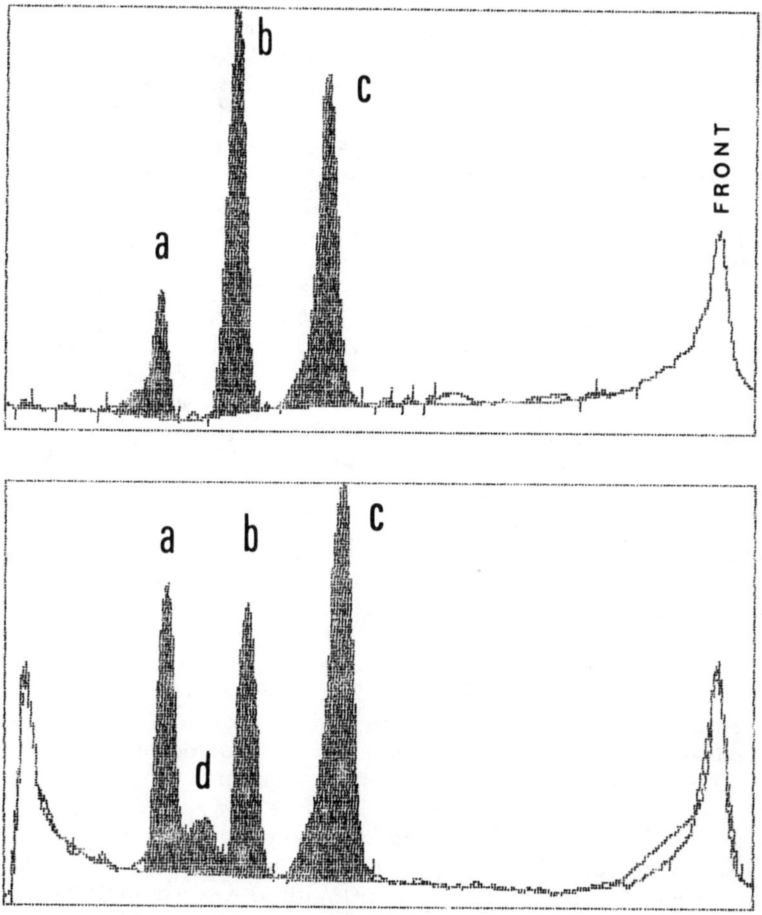

Figure 4 (A) Densitogram of standards: a, silicristin; b, silidianin; c, silybin. (B) Densitogram of a silymarin sample; d, taxifolin. The content of silymarin flavonolignanes is evaluated as siliybin.

D. Frangula

Anthraquinone laxatives have been used for thousands of years. Drugs available from senna, cascara, frangula (*Rhamnus frangula* L.), aloe, and rhubarb are well known and described in a great number of modern pharmacopoeias.

Frangula bark is one of the most frequently used plant laxatives. The fresh bark contains mainly reduced forms of anthraquinones, the anthrones. The main active principle is emodin-anthrone-6-O-α-L-rhamnosyl-8-O-β-D-glucoside. This substance has high laxative activity, but can easily cause nausea and spasms. Therefore, it is advisable to store the freshly dried drug for a year before its use. In this period anthrones oxidize to anthraquinones. In a correctly prepared drug, emodin-6-O-α-L-rhamnosyl-8-O-β-D-glucoside (glucofrangulin A) and emodin-6-O-α-L-apiosyl-8-O-β-D-glucoside (glucofrangulin B) are the main anthraquinone components. Compared with anthrones, the oxidized forms exhibit relatively mild laxative activity. In addition to glucofrangulin A and B, substances without a glucose moiety, emodin-6-O-α-L-rhamnoside (frangulin A) and emodin-6-O-α-L-apioside (frangulin B) and other minor antraquinone compounds can be detected.

Crude drugs and their extracts usually contain mixtures of active substances. These may exert a higher biological activity than would be expected from chemical analysis. This effect is due to synergism. The chemical characterization of such crude drugs is very difficult, and it usually corresponds to the biological activity of the drug. Frangula bark is an example of such a drug. As glucofrangulin A and B are the major components of frangula extracts, determination of these substances can give information about the quality of the extracts or of the drug.

We analyzed a number of ethanol-water extracts obtained under different conditions in an ultrasonic extractor and dry extracts prepared by different methods. The samples were diluted with the extraction solvent. The dry extracts were dissolved in 70% ethanol so that the final concentration of each was 20 g/l.

The separation was performed on silica gel plates (E. Merck, No. 5721). The plates were developed in the solvent system ethyl acetate-methanol-distilled water (100:13.5:10, v/v). On each plate 15 samples and 3 standards of glucofranguline A were applied. The spots were measured in the reflectance mode at 280 nm (Figure 5).

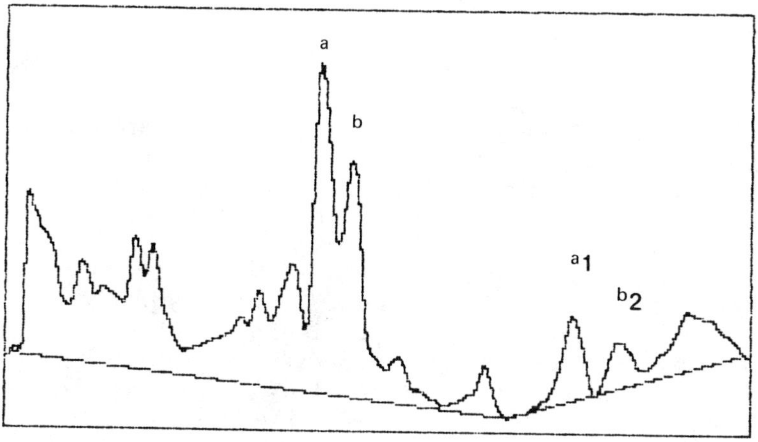

Figure 5 Densitogram of a frangula extract. a, glucofrangulin A; b, glucofrangulin B; a(1), frangulin A; b(1), frangulin B. The extract was evaluated for glucofrangulin A and B.

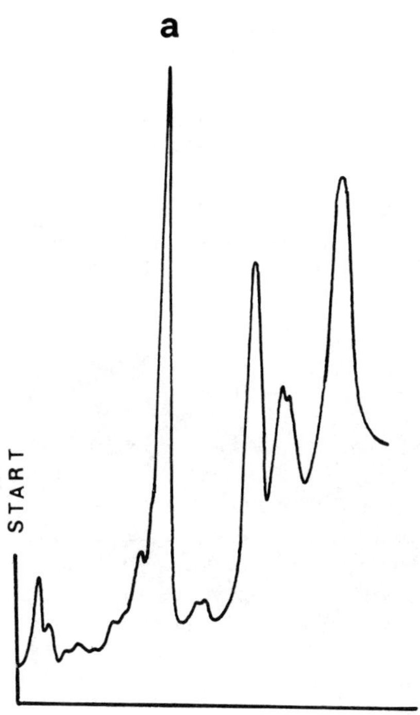

Figure 6 Densitogram of a licorice extract. a, Glycyrrhizin.

E. Licorice

Licorice, *Glycyrrhiza glabra* L. and its varieties (fam. *Leguminosae*), is another well-known medicinal plant. One of the main components of the roots is the sweet-tasting saponin glycyrrhizin (glycyrrhizic acid). On hydrolysis, glycyrrhizin yields two molecules of *D*-glucuronic acid and one molecule of an olenane type triterpene glycyrrhetinic acid.

Most of the pharmacological properties of licorice are due to the presence of glycyrrhizin, whose action resembles that of mineral- and glucocorticosteroids. This is the explanation for most of its biological effects. Its antigastric ulcer activity cannot be explained by the anti-inflammatory activity of glycyrrhizin. Furthermore, glycyrrhizin can cause side effects such as edema and hypertension. It is clear that other substances are responsible for the antigastric ulcer activity of the drug. It is possible that the main flavonone glycoside in the roots, liquiritin, or more exactly their aglycone flavonone liquiritigenin and isomeric halcone isoliquiritigenin influence this effect. Besides the anti-inflammatory and desoxycorticosterone-like actions of glycyrrhizin, the antiallergic and antihepatotoxic activities should be mentioned.

For an interesting natural substance such as glycyrrhizin, numerous different analytical methods including TLC-densitometry (13) have been developed. We used the same procedure used in the case of frangula. Samples of ethanol-water extracts were diluted with the extraction solvent and applied on TLC plates. Silica gel plates (silica gel with conc. zone, E. Merck, No. 13748) were used. The developing solvent system was *n*-butyl alcohol-acetic acid-distilled water (7:1:1, v/v). Spots of glycyrrhizin in samples and in standards were measured in the reflectance mode at 257 nm. Figure 6 shows a densitogram of a sample.

III. CONCLUSION

With these few examples of our work we intended to show the possibilities for quantitative determination of active components in extracts from different plants. Many substances from the plant kingdom have an important influence on human life. The development of plant exploitation requires modern analytical techniques. Industrial production can be economical only under defined conditions. In our production we have used TLC as the method for screening and for process control.

REFERENCES

1. U. Nyman and J. G. Bruhm. Planta Med. *35*, 97 (1979).
2. Y. Aynehchi and S. Jaffarian. J. Nat. Prod. *36*, 427 (1973).
3. M. Prosek, E. Kucan, M. Katic, and M. Bano. Chromatographia *10*, 147 (1977).
4. Y. Kiso, Y. Suzuki, N. Watanabe, Y. Oshima, and H. Hikino. Planta Med. *49*, 185 (1983).
5. H. Bernhard. Pharm. Acta Helv. *57*, 179 (1982).
6. A. K. Picman, R. L. Ranieri, G. H. N. Towers, and J. Lam. Chromatogr. *189*, 187 (1980).
7. H. Wagner, S. Bladt, and E. M. Zgainski, *Drogenanalyse*, p. 142, Springer-Verlag, Berlin, 1983.
8. H. Hikino, Y. Kiso, H. Wagner, and M. Fiebig, Planta Med. *50*, 248 (1984).
9. H. Wagner, P. Diesel, and M. Seitz, Arzneim. Forsch. *24*, 466 (1974).
10. G. Vogel. In: *Flavonoids and Bioflavonoids*, L. Farkas, M. Gábor, F. Kállay, and H. Wagner (eds.), p. 478, Akadémia Kiadó, Budapest, 1981 (Studies in Organic Chemistry, vol. 11, Elsevier, Amsterdam, 1982).
11. J. Sonnenbichler, J. Mattesberger, F. Machiao, H. Rosen, and I. Zetel. In: *Flavonoids and Bioflavonoids*, L. Farkas, M. Gábor, F. Kállay, and H. Wagner (eds.), p. 478, Akadémiai Kiado, Budapest, 1981 (Studies in Organic Chemistry, vol. 11, Elsevier, Amsterdam, 1982).
12. G. Tittel and H. Wagner. In: *Flavonoids and Bioflavonoids*, L. Farkas, M. Gábor, F. Kállay, and H. Wagner (eds.), p. 302, Akadémiai Kiadó, Budapest, 1981 (Studies in Organic Chemistry, vol. 11, Elsevier, Amsterdam, 1982).
13. Y. Takino, M. Koshioka, M. Shiokawa, Y. Ishii, S. Maruyama, M. Higashino, and T. Hayashi, Planta Med. *36*, 74 (1979).

11

Planar Chromatographic Separation and Characterization of Polymers

DANIEL W. ARMSTRONG *Department of Chemistry and Biochemistry, Texas Tech University, Lubbock, Texas*

I. INTRODUCTION

The liquid chromatographic separation of polymers has an extensive and varied history. Today the predominant polymer separation technique is that of automated column size exclusion chromatography. There has also been a recent resurgence of interest in "nonexclusion" techniques, particularly on various bonded-phase chromatographic supports (1). Much of this work was predated by thin-layer chromatographic (TLC) studies on the separation of macromolecules. Early on, planar chromatographic methods were shown to be useful in fractionating polymers. In 1959 Langford and Vaughn (2) effectively demonstrated the separation of five different polymers on a cellulose stationary phase. Both one- and two-dimensional chromatograms were run. In 1964 the first molecular weight-dependent polymer fractionation using silica gel-based TLC was reported by Spell and Eddy (3) for two different epoxy resins. From the late 1960s through the mid-1970s a considerable amount of research was done on the TLC of polymers with normal phase supports (mainly silica gel). Most of this work was done in research groups lead by Inagaki (Japan), Belenkii and Gankina (Russia), and Otocka (United States). This and the associated work of a few other groups has been reviewed a number of times (4–7).

Despite this extensive body of interesting research, TLC remained a much underutilized and often unnoticed technique for polymer analysis, particularly when compared to the increasingly popular high-performance size exclusion methods. Indeed, the TLC of polymers had essentially stagnated by the late 1970s. There were a number of reasons for this. First, most of the work had been done via normal phase planar chromatographic methods. Often complex ternary solvent systems and gradient elution (which can be difficult and can suffer from lack of reproducibility in TLC) were required. Resolution was sometimes poor due to streaking. Quantitation was not particularly good at that time. Last and perhaps most important, the mechanistic and theoretical aspects of these separations were not well understood. In general, a combination of mechanisms including adsorption, partition, phase separation, and size exclusion was responsible for the observed separations. However, it was found that one could accentuate one mechanism over the others by carefully controlling the chromatographic conditions (in particular the mobile phase composition and/or pretreatment of the normal phase plate) (4–7). Unfortunately, the interfering effects were generally not completely eliminated in the reported experiments. Clearly, a good deal of additional work could be done to optimize and further improve many of these normal phase techniques (see Sections II.B to II.J). The potential of these methods is apparent from the existing work, however.

The use of reverse phase TLC allowed one to eliminate or control many of the strong adsorptive effects which predominated in the earlier work (8, 9). This allowed the development and testing of more ideal theories and mechanisms (1, 10, 11). In general, strong, nearly ideal molecular weight dependences for a variety of polymers were easily observed on reversed phase media (8–11). Conversely, fractionating polymers by composition rather than molecular weight is more easily achieved on strongly adsorbing media such as silica gel (1–4).

Other more recent and less known work in this field includes the TLC of proteins and the separation and identification of lower molecular weight impurities from polymeric matrices. These studies as well as the reversed phase and theoretical work will be emphasized in this chapter. The earlier work (e.g., Sections II.B through II.J) will be briefly summarized to provide background for those unfamiliar with this area and to provide continuity. In all cases the reader is advised to consult original references for further details in any area of interest.

II. QUALITATIVE ANALYSIS OF POLYMERS

A. Separation and Identification of Different Homopolymers

As mentioned in the Introduction, Langford and Vaughn's successful paper chromatographic separation of polyvinyl acetate, cellulose acetate, polyvinyl butyral, polystyrene, and polyvinyl chloride was one of the earliest reported polymer separations (2). Methyl isobutyl ketone (MIBK) was used as the mobile phase in conjunction with a cellulose stationary phase. MIBK/methanol mixtures were also used in two-dimensional separations.

Simunic and Turina (12) reported a pyrolysis-TLC method for identifying a variety of polymers including polyamide-6, polyesters, melamine-formaldehyde resins, phenol-formaldehyde resins, Teflon, polystyrene, polyvinyl chloride, and high-density polyethylene. The samples were pyrolyzed for 10 min at 350°C followed by normal phase separation of the degradation products.

Tanaka et al. (13) demosntrated the effective separation of deuterated and hydrogenous polystyrenes on silica gel TLC plates. By carefully choosing the mobile phase (which affected the mech-

Table 1 Reversed Phase TLC Behavior of Polymers in Single-Solvent Mobile Phase[a]

		R_f							
Solvent	Dielectric constant	Poly-(styrene)	Poly-(α-methyl-styrene)	Polyvinyl-pyridine	Poly-isoprene	Poly-butadiene	PEG[b]	PMMA[c]	PVC[d]
Acetonitrile	37.5	0	0	0	0	0	0	0	0
Methanol	32.7	0	0	1	0	0	0	0	0
Ethylene glycol	—	0	0	0	0	0	e	0	0
Acetone	20.7	0	0	0	0	0	0	1	1
Methyl ethyl ketone	18.5	1	0	0	0	0	0	1	0
Methylene chloride	8.93	1	1	0	1	1	0	0	0

Tetrahydro-furan	7.58	1	1	0	1	1	0	1	0
Ethyl acetate	6.02	1	0	0	0	0	0	1	0
Toluene	2.38	1	1	0	1	1	0	0	1
Dioxane	2.25	1	1	0	0	0	e	1	1
Carbon tetra-chloride	2.24	1	1	0	1	1	0	1	0
Cyclohexane	1.97	0	0	0	0	1	0	0	0

[a] Data from Ref. 16.
[b] Polyethylene glycol.
[c] Poly(methyl methacrylate).
[d] Polyvinyl chloride.
[e] Severe streaking.

anism of separation) the deuterated polymer could be eluted either before or after its hydrogenous analog. For example, when the mobile phase consisted of 2-butanone-acetone (45:55, by volume) the deuterated polymer had the lower R_f value; however, when using a mobile phase that accentuated the adsorption mechanism (e.g., cyclohexane-benzene, 36:68, by volume) the hydrogenous polymer had the lower R_f. These results indicated that there are significant thermodynamic differences between isotopically different polystyrenes.

Glöckner and Meissner (14) reported the TLC separation of poly(butyl methacrylate), cellulose acetate, poly(methyl methacrylate), as well as copolymers.

Single-solvent TLC separation of polymers was first considered by Kotaka and White (15). Silica gel plates with a gypsum binder were used to separate poly(methyl methacrylate), polystyrene, and polyisobutylene. With any given solvent the polymers would either remain at the origin or move with the solvent front. Bui (16) expanded this technique to eight different polymers on reversed phase TLC plates. He found the reversed phase to be particularly effective as it eliminated problems caused by strongly adsorbing stationary phases such as silica gel. Some of these results are reproduced in Table 1. Once again, the partition coefficients of polymers in single-solvent systems tend to be either very large or very small (16). Indeed, it is apparent from Table 1 that a highly efficient qualitative analysis scheme for polymers could be devised with this technique. There is almost always one solvent that will completely separate one polymer from any other polymer. Bui (16) pointed out that this technique will separate only two polymers at a time unless a stepwise development process is used. A typical stepwise chromatogram is shown in Figure 1, where different polymer species are eluted to different levels on the same plate via multiple development. It is apparent that in single-solvent TLC of polymers, the solute will move only if (a) it is soluble in the solvent and (b) the solvent molecules interact with the stationary phase to a greater extent than the polymer (15, 16). Bui (16) also noted that in a few cases polymers of very low molecular weight would elute with the solvent front while larger polymers of the same type would remain at the origin. This was the result of a molecular weight-dependent solubility effect.

B. Separation of Different Composition Random Copolymers

Some of the earlier systematic studies of polymer TLC involved its use in the determination of compositional heterogeneities of

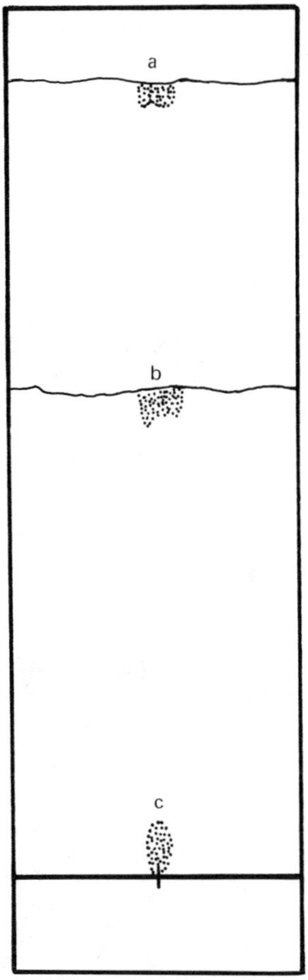

Figure 1 Stepwise separation of (A) polystyrene, (B) poly(α-methylstyrene) and (C) polyvinylpyridine on a reversed phase C_{18} TLC plate. The first development was with methyl ethyl ketone and the second development was with methylene chloride (16).

random copolymers. Ideally, chromatographic retention in this technique should be a function of chemical composition, and interferences due to other factors (such as molecular weight) should be minimal. Adsorption TLC seemed to be particularly well suited for this type of separation.

Inagaki and co-workers (17) developed the first successful TLC method for the determination of compositional heterogeneity in styrene-acrylate copolymers. Shortly afterward, Belenkii and Gankina (18, 19) reported an analogous study on styrene-methyl methacrylate polymers. The technique of Inagaki et al required a concentration gradient with a carbon tetrachloride-methyl acetate mobile phase and a silica gel stationary phase (17). The gradient was formed by addition of a second solvent during development in a vacuum desiccator. Styrene-acrylate copolymers containing between 20 and 100% styrene tended to separate by composition with little molecular weight dependence in this system. A calibration curve could be made correlating R_f to mole percent styrene composition. Inagaki et al. also reported that development with single solvents or nongradient mixed solvents generally resulted in complete retention or elution of the copolymer (17). These results are analogous to those reported for homopolymers (Section II.A). Belenkii and Gankina (18, 19) also used binary solvent mixtures (generally a chlorinated hydrocarbon such as chloroform and a polar hydrogen-bonding solvent) to achieve their separation. They found that interferences due to molecular weight were more significant for the lower MW polymers (e.g., $<10^5$ daltons).

Kamide et al. (20) found that methylene choride-alcohol mobile phases could be used to determine the acetyl content of acetylated cellulose. Molecular weight dependences were determined as well. It was found that the alcohol content had to be controlled exactly to produce optimum results. The fractionation seemed to be dependent on solvent demixing during development.

Kotaka and White (15) and Tagata and Homa (21) both studied the compositional heterogeneity of styrene-butadiene copolymers by TLC. The former workers used a carbon tetrachloride-chloroform mobile phase while the latter employed a cyclohexene-benzene binary mobile phase. Gradient elution on silica gel media was required in both cases. Once again, the relevant separation occurred over a very narrow range of solvent composition. Mori and Takenchi (22) developed a rather complex TLC scheme for the separation of copolyamides by composition. Little else has been reported on this system, however. The heterogeneity of cellulose nitrate was also determined via gradient TLC on silica gel plates by Kamide et al. (23).

Adsorption TLC and liquid chromatography (LC) on silica gel were used by Teramachi and Esaki (24) to fractionate different styrene-acrylonitrile copolymers. Methyl acetate-chloroform mobile phases effectively separated three out of four copolymers by composition. No external gradient-forming device was needed. Again, fractionation required at least a binary solvent, as single solvents either left the polymers where they were spotted or moved them with the solvent front. The TLC technique was more successful than column LC for determination of composition (24). Subsequently, Min (25) fractionated styrene-acrylonitrile copolymers on silica gel, using a tetrachloroethane-tetrahydrofuran gradient system. By using flame ionization detection, a correlation between composition and retention was demonstrated. More recently, Gankina et al. (26) demonstrated the separation of styrene-acrylonitrile copolymers (from 12 to 44% AN) by composition. Gradient elution (from carbon tetrachloride-methylene chloride mixtures to acetonitrile) was used with a silica gel stationary phase.

The separation of vinyl chloride-vinyl acetate copolymers by chemical composition was demonstrated by Janca et al. (27). Both gradient and isocratic TLC systems were developed for silica gel media. Gradient elution with methyl ethyl ketone-1,2-dichloroethane could be used to fractionate copolymers with vinyl acetate compositions between 0 and 28%. Isocratic elution could not be used for as large a composition range. For example, 1,2-dichloroethane-tetrahydrofuran-*n*-heptane (100:3:30) could differentiate between copolymers containing 6–15% vinyl acetate, while 1,2-dichloroethane-tetrahydrofuran (100:5.5) was required for copolymers containing between 10 and 28% vinyl acetate. Again it was noted that with single-solvent mobile phases, polymers either eluted with the solvent front or remained at the origin (26).

Perhaps the best example of the separation of a copolymer by composition was that by Langhorst (28) for styrene-acrylonitrile copolymers. An adsorption high-performance TLC procedure was developed for determining compositional distributions for copolymers containing from 0 to 51% acrylonitrile. Furthermore, the technique seemed to be independent of molecular weight for copolymers between 200 and 411,000 daltons. Langhorst used gradient development of a 10 × 10 cm high-performance silica gel plate in a horizontal linear development chamber. The initial mobile phase was 60:30:10, by volume, heptane-methylene chloride-acetonitrile. The second solvent (60:40, methylene chloride-acetonitrile) was added at 120 μl/min with a syringe pump (28). Evaporation from the plate was allowed to occur during development. The resulting separation is shown in Figure 2 after scanning densitometry in the dual-wavelength mode. It was noted that the calibration relationship between retention position and copolymer composition fits a second-degree polynomial (see Figure 3). The relative standard deviation was 1–4% for sample compositions in the indicated range (28). Langhorst and co-workers compared the TLC results to analogous high-performance liquid chromatography (HPLC) results.

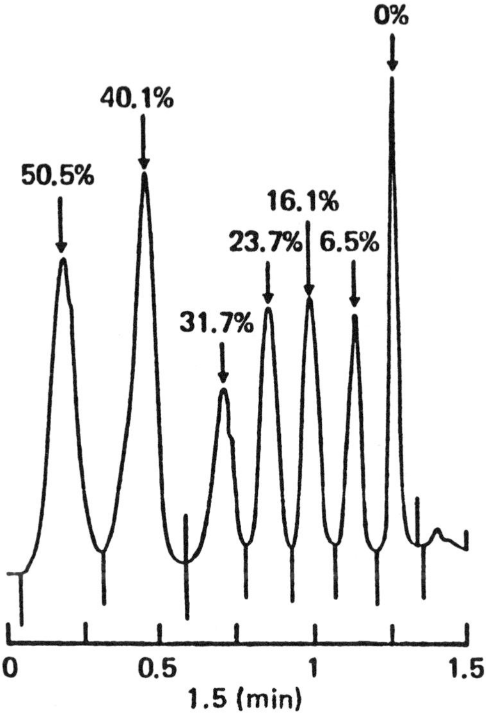

Figure 2 Normal phase, high-performance TLC separation of styrene-acrylonitrile copolymer standards according to composition. Densitometric scan (20 mm/min) was done in the UV reflectance mode (λ_1 = 263 nm, λ_2 = 360 nm). (Reprinted with permission from Langhorst, Ref. 28.)

It was thought that the TLC method was as good as or better than the simultaneously developed HPLC technique (28).

Thin-layer chromatography would seem to be an ideal technique for the determination of compositional heterogeneities of copolymers. There are, however, a few problems. In most of the aforementioned reports multiple mechanisms seem to be at work. For example, the molecular weight dependence can be minimized but generally not eliminated. Often there is a polymer spot left at the origin. In addition to the preferred adsorption mechanism, there is evidence of solubility and molecular sieving effects. Some of the earlier gradient techniques are tedious and marginally reproducible. It is likely that in the isocratic separations an "internal gradient" is

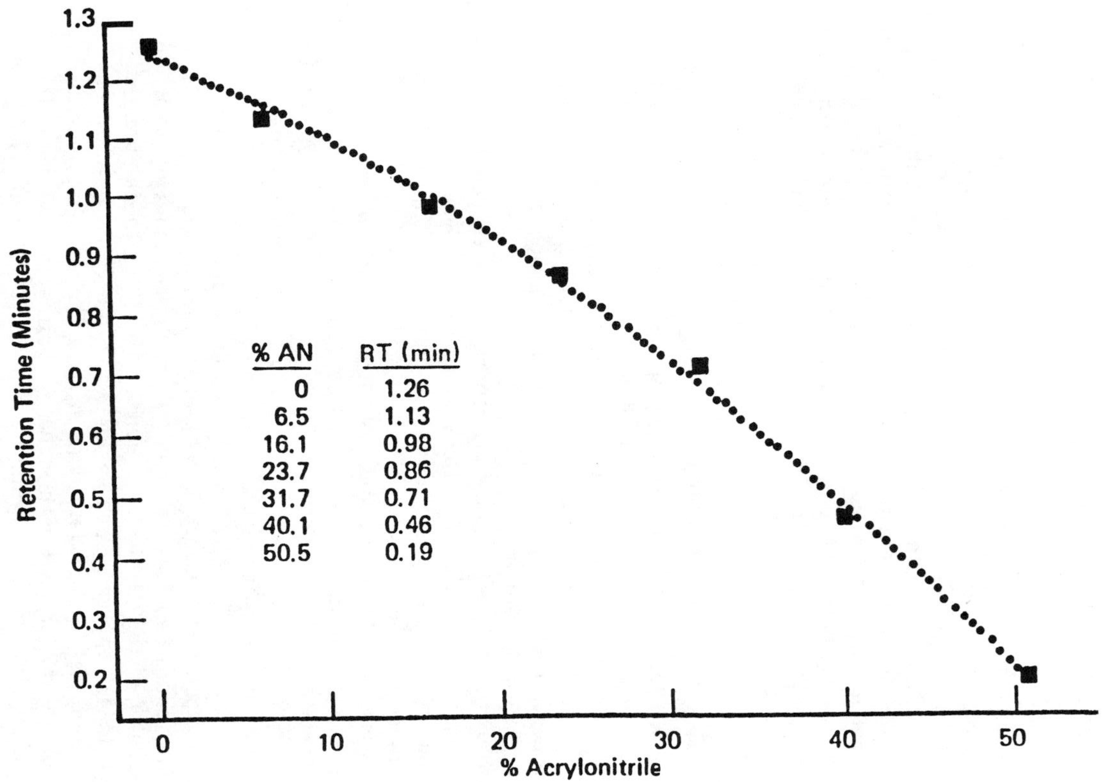

Figure 3 Relationship between spot retention and weight fraction (%) acrylonitrile in styrene-acrylonitrile copolymers. (Reprinted with permission from Langhorst, Ref. 28).

being produced by selective adsorption of one component of the mobile phase and/or by selective evaporation or equilibration of the vapor phase. In short, this is a technique of great promise but there is still considerable room for experimental and theoretical improvement (see Section V). Recently, these techniques have been extended to HPLC by Glöckner (29) and scientists at Dow Chemical (28). While the HPLC version has not been studied as extensively and can even produce poorer results, it is sometimes more amenable to theoretical and mechanistic studies. It is hoped that the use of both TLC and LC in related polymer studies will have a synergistic effect resulting in substantial improvements in both techniques.

C. Separation of Block Copolymers

The separation of block copolymers falls into two main categories. The first is the separation of copolymers by chemical composition. This is completely analogous to the separation of random copolymers described in the preceding section. The second category involves separation of copolymers of similar composition by the difference in block sequences.

One of the earliest reports on separating block copolymers according to compositional heterogeneity was by Kamiyama et al. (30) for styrene-methyl methacrylate polymers. Three different gradient solvent systems were tested on silica gel plates. Subsequent work was done by Kotaka et al. (31) and Inagaki et al. (32) on the experimental and theoretical separation of styrene-methyl methacrylate block copolymers. It was thought that these particular separations were relatively free from interferences due to molecular weight differences.

In the course of their TLC studies on compositional heterogeneities in block copolymers, Kotaka et al. and Inagaki et al. found that TLC could separate styrene (S)-methyl methacrylate (M) copolymers of different block sequence (31, 32). In particular, the diblock (SM) copolymer was easily separated from the triblock (MSM) copolymer. Gradient elution from carbon tetrachloride to 2-butanone was utilized with silica gel plates. In all cases the triblock copolymers had higher R_f values. Donkai et al. (33) extended this technique to styrene (S)-butadiene (B) block copolymers. Again, gradient elution on normal phase supports was utilized. Interestingly, the diblock SB copolymer was eluted before the SBS copolymer. No explanation was given for this reversal in elution behavior (4). Kamiyama et al. (30) pointed out an important difficulty in studying block copolymers of similar composition by TLC. They found that there was a large molecular weight dependence of the R_f for MSM block copolymers. This could lead to anomalous retention behaviors and difficulties in interpreting results.

In related studies, TLC on silica gel plates has been shown to be particularly efficient in separating homopolymer impurities from block copolymers. Specifically, the separation of polystyrene and poly(methyl methacrylate) from the block copolymer was reported (32, 33).

Again, it is apparent that TLC is a potentially powerful technique for characterizing copolymers. However, a number of problems remain, such as the lack of extensive systematic studies (perhaps due to a lack of appropriate standards), the relative lack of theoretical and mechanistic explanations for the observed chromatographic behavior, and the apparent existence of simultaneous competing mechanisms which complicate interpretations. It would seem that this is an area of great opportunity for potential investigators.

D. Separation of Graft Copolymers

Two TLC techniques have been reported which involve graft copolymers. The first involves separation of homopolymers corresponding to the main "backbone" and the side chain or graft from the grafted copolymer product. This type of separation is useful for: (a) estimating the extent of grafting, (b) evaluating solvent extraction systems (for removing the homopolymers), (c) developing appropriate conditions for preparative separations, and (d) determining the end point of the preparative solvent extraction of a reaction mixture. The second technique involves degradation of a purified graft copolymer followed by separation and quantitation to determine the true graft efficiency.

Belenkii and Gankina (19, 34) separated polystyrene (side chain) and poly(methyl methacrylate) (main chain) from the corresponding graft copolymer by using silica gel TLC plates and three different binary (and/or ternary) solvent systems. Horii et al. (35) developed TLC methods for separating the attendant homopolymers from the following five graft copolymers: polystyrene on triacetylcellulose, polystyrene on polyvinyl acetate, polyvinyl acetate on polyamide, polyethylene terephthalate on polyamide, and poly(methyl methacrylate) on polyvinyl acetate. Silica gel plates were used in all cases. Inagaki and co-workers (4, 36) also reported the separation and purification of polystyrene-cellulose and polystyrene-butadiene graft copolymers via isocratic separation on silica gel.

Taga and Inagaki's (36) work on the polystyrene-cellulose graft copolymer was particularly interesting because of the solubility problems associated with cellulose-based polymers (4, 32). It was found (by TLC) that after acid degradation of the cellulose backbone of the supposedly purified copolymer, two types of polystyrene were present. The first consisted of the polystyrene side chains which contained a terminal sugar moiety. The second was a pure homopolystyrene with no end group. The polymer containing no end group was apparently occluded during the purification step (4, 32, 36).

E. Separation of Alternating, Random, and Block Copolymers

Inagaki (4) pointed out that any comonomer pair can be polymerized to three different chain architectures, that is, alternating, random, and block copolymers. The TLC separation of these three polymer types, which contained equimolar subunit concentrations, was first considered by Kamiyama et al. (37). Both styrene-methacrylate and styrene-methyl acrylate polymers were considered. By using gradient TLC (chloroform to ethyl acetate) and silica gel plates, it was found that the styrene-methyl methacrylate block copolymer tended not to migrate compared to the other copolymers. More complex gradients could be used to separate the alternating (which had a smaller R_f) and random copolymer. A model for the separation process was also discussed (37).

Donkai et al. (33) considered the separation of different styrene-butadiene copolymers. An isocratic TLC separation (hexane-chloroform, 140:75, by volume) on silica gel media was performed. They found that the random copolymer gave the highest R_f and the block copolymer gave the lowest. Although the composition of the copolymers was slightly different, additional experiments demonstrated that it played a small role in this particular separation.

F. Separation of Linear, Branched, and Star Polymers

There have been only a few reports on the TLC separation of linear from branched and/or star polymers. Belenkii et al. (38) studied the separation of linear and branched polystyrene on silica gel plates with ternary solvent mixtures of cyclohexane-benzene-acetone. It was found that when the volume ratio of cyclohexane to benzene was held constant at 12:4, there was a pronounced dependence of R_f on acetone concentration (which was varied from 0.4 to 1.4 parts). More important, this dependence was more pronounced for the linear polymer than for the branched polymer. They thought that adsorption TLC effectively separated linear and branched-chain polystyrenes of the same hydrodynamic size (38). Further work by Belenkii and Gankina (5) demonstrated tow advantages in using combined GPC and TLC techniques. The combined procedures can be used to determine the relative amounts of linear and branched components in a polystyrene sample as well as determining their molecular weights. They were also able to determine the degree of branching in some cases (5).

Miyamoto et al. (39) and Inagaki (4) considered the separation of linear and tetrachain star polystyrenes. Star polymers are formed by reacting a linear polymer and a coupler such as 1,2,4,5-tetrachloromethylbenzene. Gradient TLC on silica gel plates easily resolved the star from the precursor linear polymer. It was found further that star polymers had higher R_f values than linear polymers of similar molecular weight (4, 39).

G. Separation of Polymers by Tacticity

Linear vinyl polymers $(—CH_2—CHR—)_n$ can have their substituents arranged in three different ways (Figure 4). These stereochemical forms are referred to as *isotactic* (where all groups are on the same side), *syndiotactic* (where the groups are arranged in alternate fashion), and *atactic* (where the groups are arranged randomly).

Inagaki et al. (40–43) first examined the TLC separation of poly(methyl methacrylate) by tacticity. It was found that an ethyl acetate mobile phase moved syndiotactic and atactic poly-(methyl methacrylate) with the solvent front while the isotactic isomer remained at the origin (40). No conclusive mechanistic evaluation of this separation has been reported.

Stereocomplexes of polymers of different tacticity form in certain solvents such as chloroform. If these mixtures are spotted on a silica gel TLC plate and developed with acetone, one can sometimes separate and identify these complexes (41–43). For example, acetone development leaves a spot of the 1:1 complex of isotactic poly(methyl methacrylate) and syndiotactic poly(methyl methacrylate) at the origin, while the atactic and excess isotactic isomers are eluted with the solvent front. Further studies were done, spotting various ratios of these polymers followed by nuclear magnetic resonance (NMR) analysis of the various separation products and complexes (41–43).

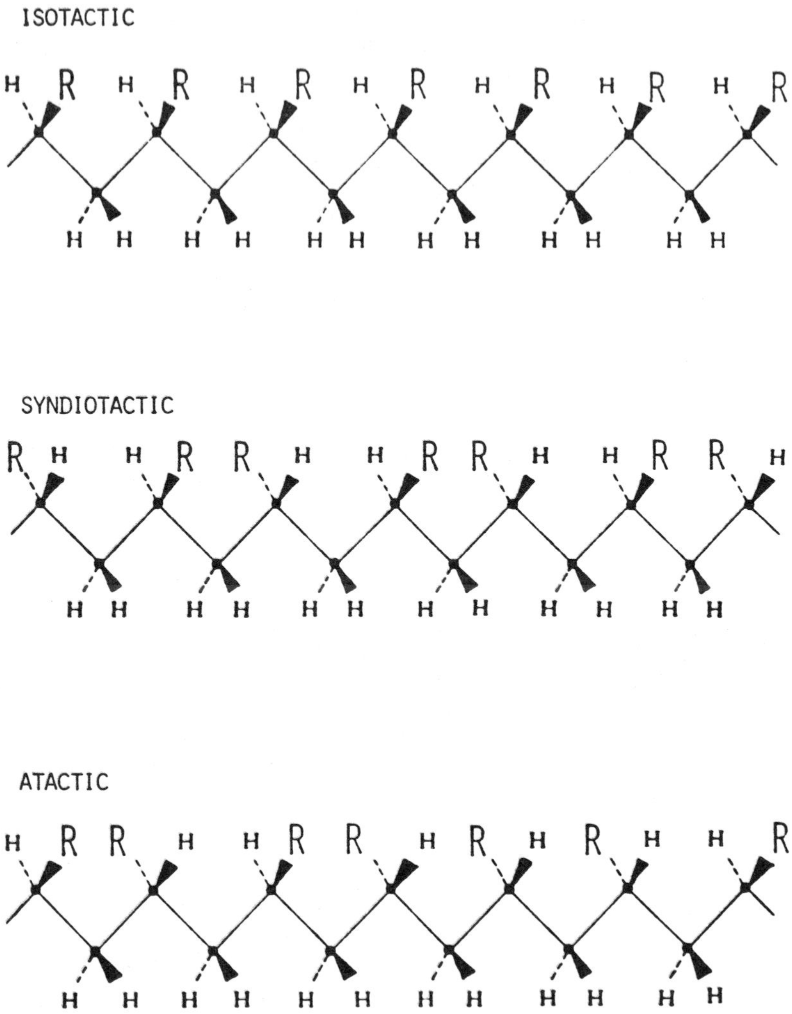

Figure 4 Schematic showing the three possible stereochemical configurations (i.e., tacticities) of vinyl polymers.

Inagaki and Kamiyama (44) reported the separation of atactic from syndiotactic poly(methyl methacrylate) on silica gel, using either an isocratic ethyl acetate-isopropyl acetate mobile phase or a gradient acetonitrile-methanol system. The polymer R_f values in this system depended not only on tacticity but also on molecular weight.

Miyamoto et al. reported that isotactic poly(methyl methacrylate) can be isolated from syndiotactic poly(methyl methacrylate) in 80–90% purity by acetone development (45). Belenkii and Gankina (5, 19) found that atactic polystyrene migrated with the solvent while leaving isotactic polystyrene at the origin when a cyclohexane-benzene-acetone mobile phase (12:4:1.5) and silica gel plates were used.

H. Separation of Polymers That Are Geometric Isomers

Polybutadiene can exist in three isomeric forms which include the *trans*-1,4, *cis*-1,4, and 1,2-vinyl analogs. Donkai et al. (46) were able to separate these via single-solvent TLC on silica gel or alumina plates. They found that development with carbon tetrachloride left the *cis*-1,4-polybutadiene at the origin while the *trans*-1,4- and 1,2-polybutadienes eluted at or near the solvent front. Conversely, when pentyl chloride was used as the mobile phase, the *trans*-1,4 isomer remained at the origin while *cis*-1,4- and 1,2-polybutadiene eluted with the solvent front. It is apparent from these results that all three polymer-isomers can be separated either by stepwise development (as described in Section II.A and Figure 1) or by two-dimensional development with the aforementioned solvents.

Inagaki extended this technique to polyisoprene (4). Elution with chloroform (on highly activated silica gel) left *trans*-1,4- and *cis*-1,4-polyisoprene near the origin while 3,4-polyisoprene moved with the solvent front. It was found that binary mobile phases of carbon tetrachloride and *p*-xylene (20:80 by volume) could be used to separate *cis*-1,4-polyisoprene (which remained at the origin) from the *trans*-1,4- and 3,4-vinyl analogs. Use of these techniques on natural rubber indicated the existence of isomeric analogs (4).

I. Separation of Polymers by End Group

Inagaki and co-workers (4, 36) have used TLC to separate polymers by the type or number of end groups present. A typical example was discussed in Section II.D on graft copolymers. Taga and Inagaki (36) were able to separate polystyrene with a terminal sugar moiety from the pure homopolymer. More recently, the separation of polybutadienes with either one or two terminal carboxylic acid or alcohol groups was reported (4). It was found that *p*-xylene development on activated silica gel eluted the pure polybutadiene with the solvent front while leaving the polymers with one or two end groups at the origin. It was also found that the monofunctional species could be separated from the difunctional species by developing with a binary solvent system (carbon tetrachloride-tetrahydrofuran, 100:1, by volume) (4). The spots were quantitated by flame ionization detection.

J. Separation of Low Molecular Weight Solutes From Polymers

Surprisingly, there have been few reports on the TLC separation of low molecular weight solutes and/or impurities from polymers even though these separations are easily achieved by TLC. Armstrong and Bui (8) showed that reversed phase TLC with a methanol mobile phase separated styrene monomers and small oligomers from all polystyrenes with molecular weights exceeding 2350. Indeed, reversed phase TLC with hydro-organic mobile phase mixtures (such as methanol-water and acetonitrile-water) can easily separate a wide variety of monomers, initiators, plasticizers, and other additives from hydrophobic polymers. This is because polymers such as polystyrene, polybutadiene, polyisoprene, and polyvinyl chloride will not move from the origin in these "poor-solvent" hydro-organic mobile phases. A typical example of the separation of styrene monomer and benzoic acid from polystyrene is shown in Figure 5.

Normal phase TLC on silica gel plates can also effectively separate low molecular weight solutes from polymers. Mazur and Lewandowski (47) reported the separation of benzophenone and benzotriazole derivatives (which are "ultraviolet stabilizers") from a vareity of commercial plastic articles. They used a cyclohexane-ethyl acetate (4:2, by volume) mobile phase. They were attempting to determine whether these low MW substances can easily find their way into foodstuffs, etc. (47). Moldovai and Moldoveanu (48) used one- and two-dimensional TLC to separate several phthalate ester plasticizers from polyvinyl chloride-based leather substitutes. Lawrence and Ducharme (49) developed a TLC method for separating fluorescent whiteners from a polymer matrix.

K. Summary

As indicated in the Introduction and the work reviewed in this section, TLC is a powerful technique for polymer qualitative analysis.

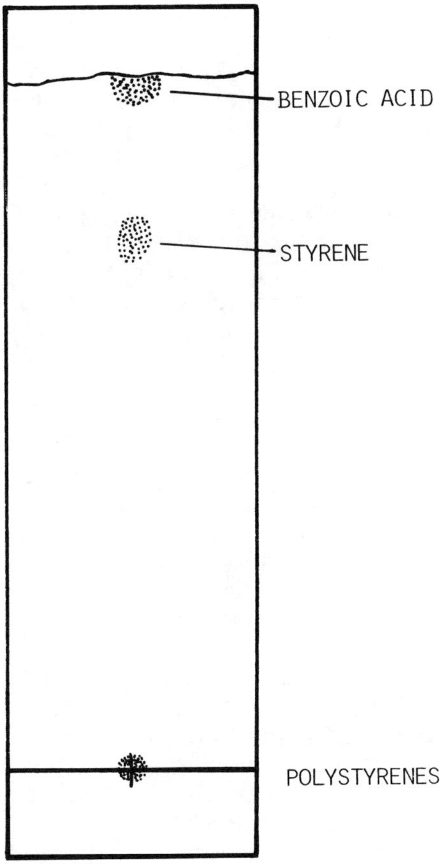

Figure 5 Reversed phase TLC separation of the low molecular weight solutes styrene and benzoic acid from polystyrene. The mobile phase was methanol and water (90:10, by volume) and the stationary phase was 10-μm C_{18}-bonded silica gel (65).

In some areas (see Sections II.D–II.J) the potential has hardly been touched. In all cases, further study and optimization is warranted and would further improve the utility of this approach to polymer analysis. Better theoretical and mechanistic understanding (Section V) will undoubtedly help to further improve all aspects of this field and lead to more ideal and powerful methods.

III. SEPARATION BY MOLECULAR WEIGHT

The ability to separate polymers by molecular weight with planar chromatographic methods is of obvious importance. However, if these methods are to be competitive with other secondary techniques for evaluating molecular weights and molecular weight distributions (in particular, high-performance gel permeation chromatography and viscosity) they must meet certain standards of practicality and reproducibility. In addition, there must be a firm theoretical and mechanistic understanding of the system. Only recently, with the advent of reversed phase technology, have these criteria been approached by thin-layer methods.

A. Size Exclusion

One of the earliest reports on the thin-layer gel permeation chromatography of polymers was by Halpaap and Klatyk in 1968 (50). By using 165 Å pore diameter silica gel and a chloroform and/or benzene mobile phase, polystyrenes with molecular weights between 5,000 and 411,000 were separated. Preelution of the plate was recommended for optimal results. As expected in a size exclusion process, the higher MW polymers were eluted first and the resolution was not particularly good. Shortly afterward, Determann (51) reviewed open-bed Sephadex systems which could be utilized for evaluating protein molecular weights. In their work on adsorption TLC, Belenkii and Gankina (19) noted the solvent-dependent appearance of molecular sieving effects. Otocka et al. (52) also studied the thin-layer GPC behavior of polystyrenes on different pore size silica gels. It was concluded that the method was of little practical value, compared to conventional TLC, because of its poor resolving power.

Perhaps the most thorough study of molecular sieve effects in TLC was done by Donkai and Inagaki (53). Using polystyrene standards with molecular weights ranging from 4,000 to 1,800,000, they evaluated the influence of stationary phase type, development procedure, preelution effects, and mobile phase effects and compared planar methods with the more common column techniques. It was apparent that size exclusion effects became significant only when adsorption effects were minimized and when the phase ratio (i.e., the weight of mobile phase per unit weight of stationary phase) remained relatively constant during development. Adsorption could be minimized by not activating the silica gel plates prior to use and by using strongly adsorbing, relatively polar solvents. The phase ratio change, which occurs during development in ordinary dry-bed TLC, can be negated by preelution of the plate with the solvent (50, 53) or by presaturation of the plate with solvent vapor (19). Donkai and Inagaki (53) were able to reproduce the size exclusion results of Halpaap and Klatyk (50) but not those of Belenkii and Gankina (19).

B. Normal Phase Fractionation

The first report of a TLC normal phase separation of polymers and oligomers by molecular weight was by Spell and Eddy (3). They were able to separate bisphenol-A-epichlorohydrin resins (from 624 to 6000 daltons) and epoxy novolacs (from 300 to 3500 daltons) on silica gel TLC plates. Two binary and two ternary solvent mobile phases were tested. In general, benzene-ethyl acetate (8:2, by volume) was most effective in fractionating epichlorohydrin resins while chloroform-diethyl ether (9:1, by volume) was preferred for the epoxy novolacs. In contrast to the aforementioned size exclusion results (Section III.A), the R_f values of the larger solutes were always less than those of their smaller analogs. It was found that different synthetic techniques produced resins with different molecular weight distributions (3). Infrared analysis of the spots allowed evaluation of the reaction mechanism as well as estimation of the extent of side reactions.

In 1970 Otocka and Hellman (54) and Kamiyama et al. (55) independently reported the fractionation of different molecular weight polymers on silica gel and alumina TLC plates. Otocka and Hellman examined the isocratic and gradient TLC separation behavior of both polystyrene (MW 10,300 to 1,800,000) and polyethylene oxide (MW 1,500 to 28,000). It was reported that isocratic separations were good over only a limited molecular weight range. Gradient elution on alumina plates seemed to produce the best separations (54). It should be noted that resolution of the moderate to high MW polymers (e.g., $\geqslant$10,000) was very poor compared to that of the smaller polymers. Kamiyama et al. also studied the fractionation of polystyrene by molecular weight on silica gel TLC plates (55). They found that elution with single-solvent mobile phases resulted in either complete elution or retention of the polystyrene standards. Only thermodynamically good solvents were able to elute the polymers, and then only if they were more strongly adsorbed on the stationary phase than the polymers. Molecular weight dependences were observed for various quaternary solvent mixtures containing benzene, acetone, butanone, and ethanol. As in Otocka and Hellman's work (54), gradient elution was found to produce the best fractionation. The molecular weight distribution of a polystyrene sample was determined by TLC and compared to that found by column GPC (55). Otocka (56) evaluated the molecular weight distribution of polystyrene by TLC with direct scanning densitometry. Again, only strong solvents were found to elute the polymers and gradient elution was found to produce the best results. Fractionation was thought to be due to solubility-based mechanisms. Although the results compared favorably with those from column GPC, the TLC method suffered from lack of consistency. It was reported that temperature, solvent purity, plate activity, chamber saturation, and mixing all affected the fractionation process (56).

Inagaki et al. (57) investigated the mechanism of molecular weight-dependent normal phase TLC. They concluded that polymers cannot be effectively fractionated by molecular weight via TLC unless adsorption and desorption processes are minimized and a precipitation mechanism is predominant. In the course of this study it was demonstrated that poly(methyl methacrylate) (from 4,000 to 41,000 daltons) could be separated by TLC techniques analogous to those developed for polystyrene (57). Additional work on the molecular weight-dependent fractionation of polystyrene, using a variety of isocratic and gradient solvent systems, was reported by Otocka et al. (58) and Taga and Inagaki (59). In 1973 Otocka (60) wrote a brief review on the determination of polymer molecular weights and distributions by normal phase TLC and scanning densitometry.

Kotaka and White (15) first studied the TLC fractionation of styrene-butadiene copolymers by molecular weight. All previous molecular weight fractionations (vide supra) were done on homopolymers. Gradient elution was required in order to see a molecular weight dependence on silica gel plates. As in the case of the homopolymers, there was a nearly linear relationship between R_f and log MW. A comparison of this TLC method with column GPC indicated that TLC gave too narrow a distribution of molecular weights (15). This was largely due to spot quantitation problems with the TLC method. Further inaccuracies may have resulted from the fact that some polymer residue tends to be left at the origin in normal phase techniques (15). Inagaki et al. (32) also considered the separation of styrene-butadiene copolymers. In particular, they discussed factors (i.e., solvent composition) that allowed one to separate copolymers by composition vs. block sequence vs. molecular weight. Obviously, a gradient-induced precipitation or solubilization mechanism is preferred for the latter while a strong adsorption mechanism is needed for the former.

Perhaps the most interesting discussion on the separation of polymers by molecular weight, with normal phase TLC, and the factors that control the mode of separation was by Belenkii and coworkers (5, 61). Both theoretical and experimental aspects of homopolymer separations were considered. The theory will be covered in Section V. It was thought that with the proper solvent combinations, adsorption TLC could be used to fractionate styrene polymers by molecular weight (5, 61). Indeed, reasonable separations of polystyrenes between ~1,000 and 2,200,000 daltons (with ternary mobile phases) were shown. It was demonstrated experimentally that an adsorption mechanism was responsible for the separation.

Obviously, there is some disagreement about the mechanism and best experimental approach for normal phase TLC fractionation of polymers by molecular weight. Some researchers have concluded that adsorption TLC cannot effectively separate polymers by molecular weight (15, 62), while others believe it is better than the precipitation-solubilization approach (5, 61). In general, the normal phase techniques suffer from a number of shortcomings. For example, they are most effective for the lower molecular weight polymers and have trouble resolving macromolecules with molecular weights over 10,000. Only a limited number of polymers have been fractionated. Complex gradient and ternary solvent systems (which can suffer from lack of reproducibility) are required. There is some question about the exact mechanism of separation. In fact, it is likely that there are several competing mechanisms at work. Lastly, there is no good theoretical framework which allows one to accurately describe and predict the separation behavior of polymers. This is not to say that reasonable, *idealized* theories have not been proposed (see Section V), only that experimental results indicate that the situation is somewhat more complicated than indicated by any present theory.

C. Reversed Phase Fractionation

The reversed phase TLC fractionation of polymers by molecular weight eliminated many of the experimental and theoretical problems associated with the normal phase methods. Armstrong and Bui (8) first studied this approach using TLC supports of octadecylsilane-bonded silica gel. As reported in previous normal phase work, single-solvent mobile phases tended to elute the polymer with the solvent front or not at all. Specifically, thermodynamically good solvents (such as methylene chloride or chloroform) eluted hydrophobic polymers of all molecular weights, while thermody-

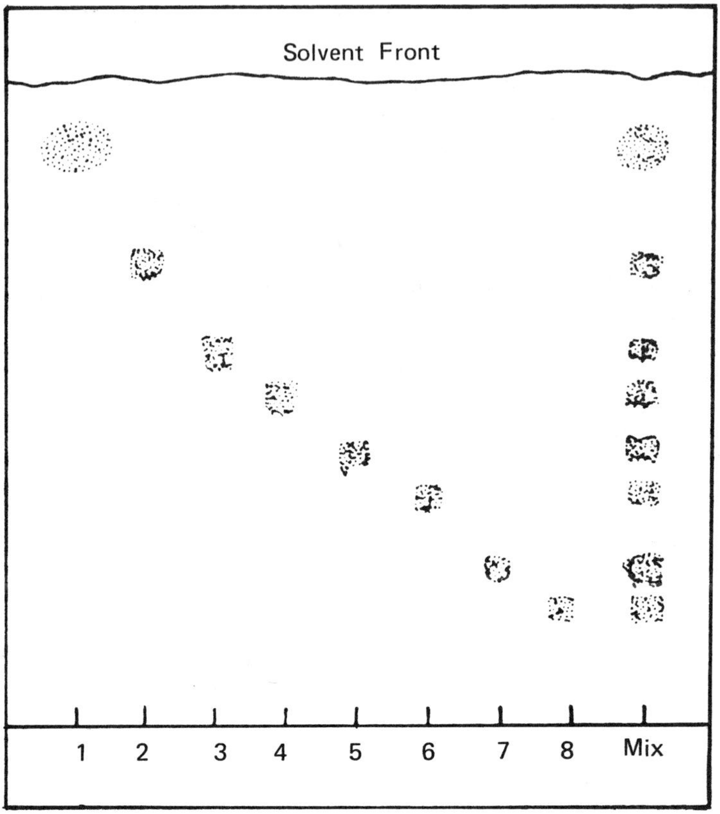

Figure 6 Schematic of the reversed phase TLC fractionation of polystyrene in an open system. Mobile phase composition was methylene chloride-methanol (79:21, by volume). Polystyrene molecular weights are: 1, 105 (monomer); 2, 2.35×10^3; 3, 1.75×10^4; 4, 3.5×10^4; 5, 1.1×10^5; 6, 3.9×10^5; 7, 3.7×10^6; and 8, 1.0×10^7 (8).

namically poor solvents (such as methanol or ethylene glycol) left all polymers at the origin. Thermodynamically poor solvents or aqueous mixtures of these solvents could be used to elute monomers and some oligomers, however. The proper binary mixture of a good and a poor solvent could be used to fractionate polystyrenes with molecular weights between 2000 and 10 million (Figure 6). No complex gradient elution of ternary or quaternary solvent systems was needed. One could "focus" the separation on any given molecular weight range by adjusting the mobile phase ratio of methylene chloride to methanol (8). The efficiency, molecular weight fractionation range, and reproducibility were superior to those of the normal phase methods. At least part of the reason for this was that the reversed phase technique eliminated most of the competing strong adsorption effects. Attempts to repeat this fractionation by isocratic reversed phase HPLC failed (8). It was found that a gradient was required for an effective LC separation. This was surprising in view of the isocratic TLC results. However, further experiments indicated that reversed phase TLC plates formed a spontaneous gradient during development with binary solvents. In the case of methylene chloride-methanol mobile phase, for example, the more hydrophobic methylene chloride was preferentially adsorbed by the stationary phase during development, thereby enriching methanol in the remaining solution (8). Interestingly, allowing a small amount of evaporation to occur accentuated the gradient and improved the separation.

It became apparent that one could easily fractionate a variety of polymers by reversed phase TLC by judicious choice of a binary solvent mobile phase. Two conditions had to be satisfied, however. First, the thermodynamically good solvent must be preferentially adsorbed by the stationary phase, thereby enriching the developing mobile phase in the thermodynamically poor solvent. Second, the proper ratio of solvents must be determined exactly. Small differences in the starting composition of the mobile phase resulted in fairly large differences in the fractionating power of this technique (8, 63). Armstrong et al. (63) and Bui (16) fractionated a variety of polymers by reversed phase TLC, including polystyrene, poly(α-methylstyrene), polyisoprene, polybutadiene, polyethylene glycol, polyethylene oxide, polytetrahydrofuran, poly(methyl methacrylate), polyvinyl chloride, polyvinyl alcohol, and polyvinyl acetate. Typical results and separation conditions are given in Table 2. There was generally a linear relationship between R_f and log MW for all but the very large polymers and some oligomers (16, 63). The theory and mechanism of separation were considered in detail (see Section V.B).

Bui (16) evaluated several different bonded-phase media (i.e., C_2, C_8, and C_{18}) to determine the optimum stationary phase for polymer fractionation. Factors considered were development time, retention, spot shape, and resolution. In general, the resolution and spot shape were best for the C_{18} bonded media. However, depending on the binary solvent mobile phase used, development times varied considerably with each medium.

Table 2 Polymers Which Have Been Fractionated by RPTLC

Polymers	Solvent pairs	Reference
Polystyrene	$MeCl_2$/MeOH	8
Poly(α-methylstyrene)	$MeCl_2$/MeOH	9
Poly(methyl methacrylate)	$MeCl_2$/MeOH; THF:EG	9
Polyethylene glycol	Dioxane/ethylene glycol; MeOH/EG	9
Polyethylene oxide	Dioxan/EG	9
Polyvinyl chloride	THF:EG	68
Polyvinyl acetate	THF:EG	68
Polyisoprene	$MeCl_2$/MeOH	9
Polybutadiene	$MeCl_2$/MeOH	9
Polytetrahydrofuran	THF:EG	9

A phase separation mechanism was responsible for the reversed phase TLC separation of polymers. Consequently, separation by molecular weight was governed not only by the composition of the mobile phase, but also by the volume of mobile phase present (or, more exactly, the phase ratio). It is well known that the phase ratio (i.e., the weight of the mobile phase over the weight of the stationary phase) decreases during TLC development. Inagaki (64) first indicated that this phenomenon might be useful in fractionating polymers. Few definitive studies were done, however. Armstrong and Bui (65) used this phenomenon to accentuate the polymer fractionating power of reversed phase TLC. Beveled plates were used, as indicated in Figure 7. The thickness of the stationary phase at the bottom of the plate was 250 μm and tapered smoothly to less than 25 μm at the top of the plate. The polymer samples were spotted near the thickest portion of the plate and developed toward the narrow end. The net result was that the amount of mobile phase available to carry the polymers was decreased to an even greater extent than observed for traditional TLC plates. Fractionating power increased two- to fivefold on these nontraditional plates (depending on molecular weight).

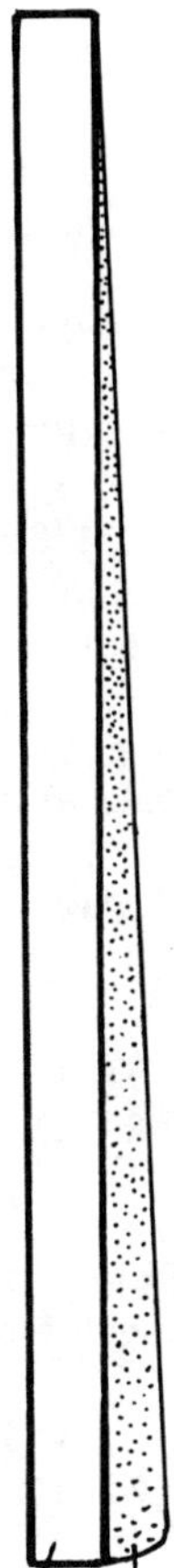

Figure 7 Side view of a beveled, reversed phase TLC plate. The thickness of the stationary phase generally varies from ∿1 mm at the bottom of the plate to 25 μm at the top. Polymer fractionation is sometimes enhanced with this type of TLC plate (65).

Bui and Armstrong (66) compared the separation efficiency and resolution of reversed phase TLC with that of high-performance GPC and gradient LC. The specific resolution (Rsp) of these techniques was calculated for different molecular weight standards of polyisoprene, polyethylene glycol, polyethylene oxide, polystyrene, and poly(α-methylstyrene) from the following equation:

$$Rsp = \frac{0.576}{\sigma D_2}$$

where σ is the standard deviation of the polymer standard peak and D_2 is the slope of the linear portion of the calibration curve (66). It was found that gradient LC has an order of magnitude greater resolving power for high polymers than GPC, while TLC had two to fives times the resolving power of GPC. This was largely due to the greater selectivity of the LC and TLC techniques. It was also noted that four separate GPC columns were needed to cover the molecular weight range of the polymers investigated (5,000 to 400,000) while only a single reversed phase C_{18} TLC plate or column was required (66).

The reversed phase TLC of polymers has proved to be so facile and effective that it has been recommended as an undergraduate organic chemistry experiment (67). In the experiment, styrene is polymerized and its average molecular weight and molecular weight distribution are followed with time, using TLC. For the first time students in large-scale teaching laboratories are able to evaluate what they have made in a polymerization experiment as well as the factors that control polymer formation (such as concentration, temperature, amount of initiation, type of initiator, and catalysts).

The use of reversed phase TLC to fractionate polymers by molecular weight eliminated most of the strong adsorption effects that complicated and limited analogous normal phase methods. This resulted in a less tedious, more efficient, and more reproducible method for fractionating polymers. Perhaps more important, this separation was sufficiently "ideal" that the theory and mechanism could be accurately evaluated. As will be seen in Section V, this is the only TLC method in which a theory derived from first principles has been used to predict experimental R_f values.

D. Determination of Molecular Weight and Molecular Weight Distributions

Thin layer chromatography is a "secondary" method for determining molecular weights and molecular weight distributions, as are viscosity and GPC or size exclusion techniques. By secondary, it is meant that one must compare one's results for an unknown polymer to those of a series of well-characterized standards. "Primary" techniques such as osmometry, light scattering, or ultracentrifugation can, in theory, be used to determine molecular

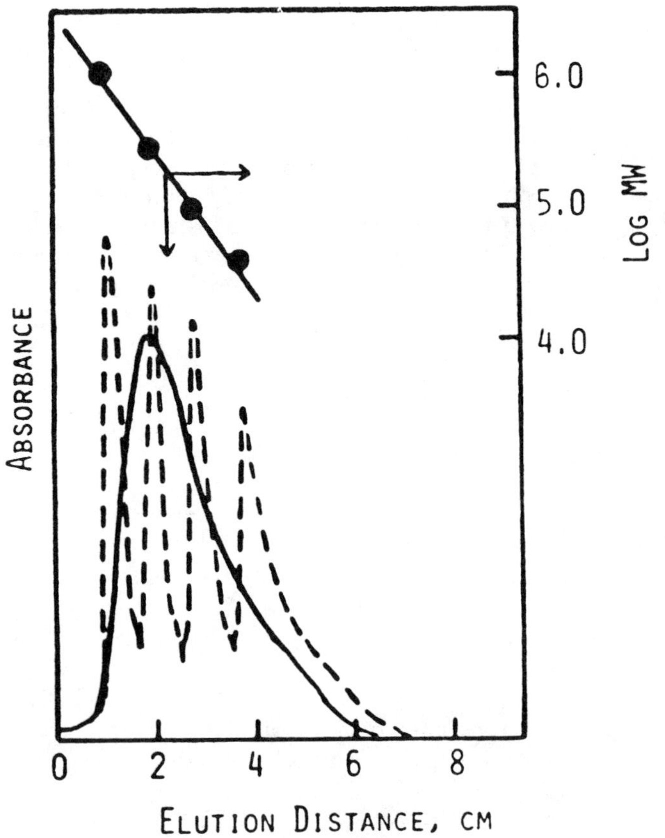

Figure 8 Two superimposed scanning densitometric profiles of a polystyrene unknown (solid line) and polystyrene standards. Above the profiles is the calibration curve for the standards (log MW vs. elution distance) (68).

weights and distributions without standards or standard curves. However, even with primary techniques it is common practice to run at least one well-characterized macromolecule to see if the instrument and operator are functioning properly.

There have been at least four reports on the use of TLC to determine the molecular weight and molecular weight distribution of polymers. Two of these, Kamiyama et al. (55) and Otocka (56), utilized normal phase TLC while the others, Bui and Armstrong (66, 67), utilized reversed phase TLC. Otocka (60) also reviewed the normal phase method. There are several steps in evaluating the molecular weight of a polymer by TLC. First, one must determine the relationship between R_f and molecular weight for a given system. This is done by separating several "narrow-disperse" standards and making a standard curve (see Figure 8). The polymer unknown is generally spotted on the same plate as the standards so that the separation conditions will be as nearly identical as possible. The R_f of the upper and lower edge of the unknown spot (or streak) defines the distribution of molecular weights of the polymer. Finally, the total amount of unknown polymer at

Table 3 Definitions of Molecular Weight Averages

A. Number average molecular weight (Mn)

$$Mn = \frac{\sum n_i M_i}{\sum n_i} \quad \text{or} \quad \frac{\sum w_i}{\sum w_i / M_i}$$

where n_i and w_i are the number of moles and the weight, respectively, of the polymer having molecular weight M_i.

B. Weight average molecular weight (Mw)

$$Mw = \frac{\sum n_i M_i^2}{\sum n_i M_i} \quad \text{or} \quad \frac{\sum w_i M_i}{\sum w_i}$$

C. z-average molecular weight (Mz)

$$Mz = \frac{\sum n_i M_i^3}{\sum n_i M_i^2} \quad \text{or} \quad \frac{\sum w_i M_i^2}{\sum w_i M_i}$$

D. The ratio of the weight average molecular weight and the number average molecular weight (Mw/Mn) is used as a measure of polymer polydispersity. The Mw/Mn of a monodisperse polymer is 1. This number increases with polymer polydispersity.

all R_f's must be determined. This is done by accurately quantiating the unknown by some densitometric method (see Figure 8). From the standard calibration curve and the densitometric profile of the unknown polymer one can calculate any desired molecular weight average including number average, weight average, and z-average molecular weights (see Table 3).

Kamiyama et al. (55) compared the number average molecular weight and molecular weight distribution (of a polystyrene standared) obtained by normal phase, gradient TLC to that obtained by GPC. No clear cut distinction was seen between the results of the two methods. Quantitation of the developed polymers was done by first visualizing the spots with thymol blue followed by 3 N H_2SO_4(aq). The chromatogram was then photographed on high-contrast film. The degree of blackness of the film was proportional to the amount of polymer present (55).

Otocka (56, 60) also compared a polystyrene standard's molecular weight and molecular weight distribution obtained by normal phase, gradient TLC with that obtained by GPC. The spots were visualized by the perchloric acid method and the "unknown" was quantitated by scanning densitometry. Again, the TLC and GPC results were analogous. It was pointed out that the successful fractionation and determination of molecular weight by normal phase, gradient TLC is something of an art (56). Temperature, solvent purity, plate activity, chamber saturation, and mixing can all significantly affect the outcome of this technique.

Isocratic reversed phase thin-layer chromatography (RPTLC) and scanning densitometry were used to determine various molecular weight averages and distributions of a broad moelcular weight range of samples of polystyrene and poly(methyl methacrylate) (68). The RPTLC results were compared to those obtained by the National Bureau of Standards using light scattering, sedimentation equilibrium, osmometry, and solvent fractionation. With properly controlled conditions, RPTLC results were accurate and reproducible (see Table 4). A basic program was developed which analyzed the analog signal from the scanner and rapidly calculated, printed, and graphed all desired results. In order to obtain consistent and accurate results by TLC, it was found that the proper amount of polymer must be spotted ($\leqslant 15$ μg) and background from the plate must be minimized (68). It was found that no correction for band broadening due to processes other than the fractionation process (eddy diffusion, mass transfer, etc.) was needed as long as proper RPTLC conditions were employed (68). In fact, inaccuracies due to band broadening seem to be much more serious in GPC and normal phase TLC than in RPTLC.

Table 4 Comparison of Polymer Molecular Weight Averages Obtained by RPTLC and Other Traditional Methods

	Average Mw by RPTLC	Standard deviation	Average MW given by manufacturer
	Polystyrene[a] (NBS)		
Mw	2.45×10^4	9000	2.58×10^4 (light scattering 2.88×10^4 (sedimentation equilibrium)
Mn	1.20×10^4	9000	1.23 (fractionation)[b]
Mw/Mn	2.0	—	2.1
	Poly(methyl methacrylate)[c] (Poly Sciences)		
Mw	8.0×10^4	3000	8.1×10^4 (light scattering)
Mn	5.4×10^4	3000	4.7×10^4 (osmometry)
Mw/Mn	1.5	—	1.7

[a]Fractionated using 78:22 (v:v) $MeCl_2$/MeOH. Sample loading, 5 mg/ml. Analyzed in the reflectance mode at 265 nm after spraying with ethanolic sulfuric acid solution.
[b]Based on fractionation value of Mw/Mn multiplied by Mn value for light scattering.
[c]Fractionated using 30:70 (v:v) ($MeCl_2$/MeOH). Sample loading, 5 mg/ml. Analyzed in the transmission mode at 405 nm after spraying with methanolic iodine solution.

A quick, less accurate RPTLC method for evaluating molecular weights and molecular weight distributions was reported by Armstrong et al. (67). In this method one can avoid quantitating the unknown polymer by assuming that its molecular weight distribution is Gaussian. This method is excellent for obtaining rapid, approximate results.

IV. SEPARATION OF PROTEINS

There have been few reports in the literature on the TLC separation of proteins. This is surprising in view of the potential usefulness of this method. Armstrong and Dinyarian (69) have completed preliminary studies in this area, the results of which are given below. Proteins were separated on both cellulose and reversed phase TLC plates. The rationale and mechanism of separation on these two phases were very different, however. Aqueous solutions of polyethylene glycol (PEG; MW 3300) were used as mobile phases with the cellulose plates. Retention data are given in Table 5 and a typical chromatogram is shown in Figure 9. Many water-soluble proteins have high R_f's on cellulose plates when 100% aqueous or buffered mobile phases are used. The addition of sufficient quantities of water-soluble polymers (such as PEG) can cause proteins to precipitate from solution. This is believed to be due to an excluded-volume effect. That is, the PEG takes up so much volume that the protein is forced out of solution. Each individual protein requires a somewhat different quantity of PEG to be present before phase separation occurs. Consequently,

Table 5 Effect of Polyethylene Glycol on the R_f Values of Several Proteins on Cellulose Plates[a]

	Protein R_f values as a function of percent polyethylene glycol in mobile phase			
Protein	3.5%	7%	15%	30%
Egg albumin	0.82	0.81	0.78	0.00
Carbonic anhydrase	0.90	0.90	0.0	—
α-Lactoalbumin	1.00	1.00	1.00	0.9
β-Lactoglobulin	0.80	0.80	0.70	0.00
Lysozyme	0.00	0.00	0.00	0.00
Phosphorylase B	1.00	1.00	0.88	0.00

[a]The polyethylene glycol (Aldrich) had an average molecular weight of 3300. Unbuffered aqueous solutions were used. Cellulose plates (K2F) were obtained from Whatman.

Figure 9 Separation of the proteins β-lactoglobulin, lysozyme, and phosphorylase B on a cellulose TLC plate with a mobile phase of 3.5% polyethylene glycol in water (69).

by controlling the amount of PEG in the mobile phase, one can separate a wide variety of these biologically important macromolecules. This is nicely demonstrated by the results in Table 5 (69). Not only did the protein R_f's decrease with increasing PEG, but each solute also seemed to undergo a rather sharp transition from low retention to high retention at a characteristic PEG concentration.

The reversed phase TLC of proteins is considerably more difficult and complicated than that of synthetic polymers. A number

Table 6 Reversed Phase TLC Separation of Proteins[a]

	Volume ratio of 0.2 M phosphate buffer (pH 4.0) to acetonitrite			
Protein	100% buffer	75/25	50/50	80/20 (pH 7.0)
Egg albumin	0.00	0.63[b]	0.20[b]	0.53
β-Lactoglobulin	0.00	0.00	0.20[b]	0.00
Lysozyme	0.00	0.15[b]	0.00	—
Phosphorylase B	0.50	—	0.00	—

[a]The TLC plates were 5 × 10 cm Whatman KC18F.
[b]Some streaking occurred.

of theoretical, mechanistic, and experimental problems exist for proteins which are not as significant for most synthetic polymers. These problems have been discussed by Armstrong and Boehm (1). Proteins tend to bind strongly and often irreversibly to silica gel-based bonded-phase media. Denaturation frequently occurs, although this is not important in the case of strictly analytical separations. In some cases elution can occur with buffered hydroorganic mobile phases. Typical results are given in Table 6. Both the volume ratio of organic modifier to buffer and the pH can affect the separation. Streaking can be a problem in some cases. Furthermore, the solvent used to spot the protein and different treatments of the spotted plate before development can cause altered retention behavior. This is presumably due to the fact that proteins are easily denatured and chemisorbed under these separation conditions.

V. THEORY AND MECHANISM

A. Normal Phase

The theoretical aspects of the normal phase separation of polymers has been considered extensively by Belenkii and Gankina (5, 6, 61) and qualitatively be Inagaki (4). As these theories have been reviewed (4–6, 61), only a brief synopsis will be given here.

The mechanisms said to be responsible for polymer separation on normal phase TLC include: (a) adsorption, (b) solubility-controlled phase separation, and (c) molecular sieving (4). Belenkii and Gankina (5) divide the solubility-controlled phase separation mechanism into precipitation and extraction TLC. Conceptually, ideal theories can be and have been formulated for all of these separation mechanisms. Unfortunately, the gulf between these ideal theories and what actually occurs on a normal phase TLC plate is sometimes considerable. In addition, the mechanistic and theoretical problems that have complicated the TLC of small moleules (the change in phase ratio, vapor effects, activation, wetting of a dry bed, etc.) are even more complex for polymers. For example, it is difficult to engineer the conditions of a normal phase TLC experiment so that a polymer separation is dictated exclusively by one ideal mechanism. Indeed, comments on the interference of "molecular weight effects" in the qualitative analysis of polymers and adsorption effects in phase separation TLC are common throughout the literature. Regardless of these difficulties, it is sometimes beneficial to consider the ideal cases in the hope of better understanding the more complex "real-life" situation.

Belenkii and coauthors developed the theoretical framework for macromolecular adsorption TLC. It can be shown that, for an ideal chromatographic system, the change in free energy of the system ($-\Delta G$) is equal to the change in free energy of the macromolecules when adsorbed (61):

$$-\frac{\Delta G}{kT} = \ln k' + \ln \frac{V_0}{V_p} \qquad [1]$$

where k' is the capacity factor, V_0 is the mobile phase volume, V_p is the stationary phase volume available for adsorption (sometimes given as pore space), k is the Boltzmann constant, and T is the absolute temperature. Based on the work of DiMarzio and Rubin (70), it can be shown that the change in free energy $-\Delta G/kT$ accompanying polymer adsorption is dependent on the interaction energy ($-\varepsilon$) of a polymer segment with the surface of the porous adsorbent. It has been shown that there is a critical energy ($-\varepsilon_{cr}$) where $-\Delta G$ is zero. At this point $-\varepsilon = -\varepsilon_{cr}$ and the entropy losses ($T\,\Delta S$) of the macromolecule completely compensate for the increase in enthalpy, ΔH (61). Consequently, there are three possible separation modes when LC is performed with a strongly adsorbing stationary phase. When $-\varepsilon > -\varepsilon_{cr}$, adsorption effects dominate and control the separation; when $-\varepsilon < -\varepsilon_{cr}$, the entrance of macromolecules into adsorbent pores is controlled by molecular sieving effects; and when $-\varepsilon = -\varepsilon_{cr}$, the above two effects cancel one another and the distribution coefficient is unity for all polymers regardless of molecular weight (5, 61).

Thin-layer precipitation chromatography of polymers can theoretically occur on any appropriate porous support when gradient elution from a thermodynamically poor to a thermodynamically good solvent system is used. The pore size of the support should be sufficiently small to prevent access to all polymers. The polymer is then restricted to the excluded volume (Ve) of the column. The average linear velocity (μ_p) of a dissolved polymer in such a system is given by

$$\mu_p = L/(Ve/v) \qquad [2]$$

where L is the bed length and v is the flow rate (1, 28).

Unlike the polymers, the mobile phase molecules can penetrate the pores and therefore have access to a greater volume, referred to as the void volume (Vm). Thus,

$$Vm = Ve + Vi \qquad [3]$$

where Vi is the imbibed volume of the packing (i.e., the internal pore volume). Consequently, the average linear velocity of the solvent (μ_s) is

$$\mu_s = L/(Vm/v) \qquad [4]$$

Because Vm is greater than Ve, the velocity of the dissolved polymer is greater than that of the solvent. If a polymer that is immobilized at the beginning of a bed is eluted with and increasingly good solvent, then the polymer will eventually dissolve and quickly move ahead to the poor-solvent region and be reprecipitated. This process can occur several times before elution (1, 29). Except for the role of the pore size, this model assumes that the stationary phase plays no role in the separation process. This is not true for normal phase TLC, where strong adsorption effects cannot be eliminated. It is also not entirely true for reversed phase TLC, as will be seen in Section V.B.

Inagaki (4) first made the distinction (in chromatography) between a simple polymer precipitation and a thermodynamic phase separation. This is a subtle but important difference. According to Inagaki, development with a solvent results in a phase equilibbrium, in which polymer molecules are in two phases, a sol and a gel phase, according to their molecular weights (4). Solutes in the sol phase are carried with the developer while those in the gel phase are strongly retarded. The molecular weight of a polymer dictates whether it exists in the sol or gel phase (the higher the MW, the more likely it is that the polymer exists in the gel phase). Thus higher molecular weight polymers would have lower R_f values than lower molecular weight polymers (9). It was realized that although this provided a simple, qualitative explanation for the separation of polymers by molecular weight, it was an idealized system.

B. Reversed Phase

Armstrong and co-workers (8–11) hoped that a well end-capped reversed phase medium would eliminate many of the problems observed with normal phase TLC. Using statistical mechanics, a phase separation chromatographic model system was developed which not only accounted for many of the phenomena observed but also could be used to predict the R_f values of polymers in TLC and/or retention times in HPLC (9–11, 16). The model system assumed an equilibrium distribution of isolated flexible polymer molecules between a binary solvent mobile phase and a planar stationary phase. In the mobile phase the polymer was assumed to be a spherical gel having a uniform segment density. On the stationary phase the polymer assumed a flat, thin cylindrical conformation. For polymers of sufficiently high molecular weight there is a critical mobile phase mole fraction, X_{1mc}, of the more favorable, less polar solvent 1 such that when $X_{1m} > X_{1mc}$ the polymer will move at the same rate as the mobile phase but when $X_{1m} < X_{1mc}$ the polymer is completely retained by the stationary phase (see Figure 10). The critical composition is dependent on molecular weight. This dependence results from the flexibility of the polymer, which enables it to change configuration in response to its environment. As a result of the X_{1mc}-molecular weight dependence, one can chromatographically fractionate homopolymers with gradient elution techniques (8). In this type of TLC fractionation, the mobile phase changes composition continuously during development as a result of selective adsorption of the better solvent by the stationary phase. To achieve fractionation by this method one must carefully choose a good-poor solvent pair that will interact differently with the stationary phase and produce the optimum gradient.

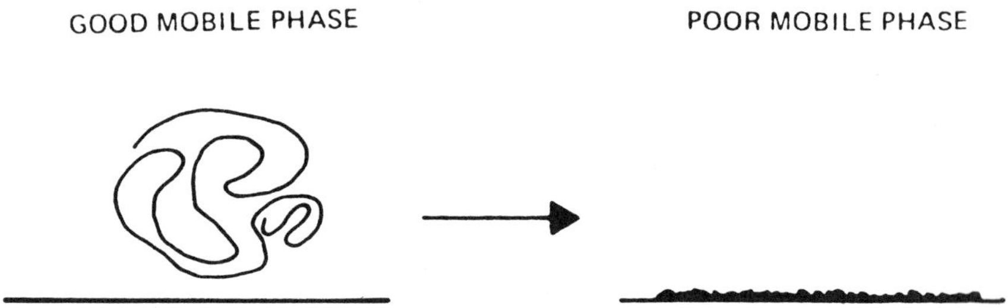

Figure 10 Illustration of the two theoretically possible configurations for a high molecular weight polymer in reversed phase TLC or LC gradient fractionation (9).

One can obtain the theoretical expression for the capacity factor of a polymer by using the Flory-Huggins lattice model for isolated polymer-solvent systems in both the mobile and stationary phases, as well as the Bragg-Williams approximation for the nearest-neighbor interactions (10, 71, 72). From the expression for the capacity factor one can obtain equations relating retention time (in LC) and R_f (in TLC) to polymer molecular weight (vide infra).

$$t_r = t_m + B^{-1}(X_{1mc} - X_{1m(0,0)}) + (|A_1|MB)^{-1}\ln(|A_1|MBt_m) \qquad [5]$$

where t_r is the retention time of the polymer, t_m the retention time of an unretained solute, X_{1mc} the critical solvent composition for the polymer, $X_{1m(0,0)}$ the initial mobile phase composition, B the rate at which the mobile phase composition varies, A_1 a constant depending on solvent-solvent and solvent-polymer interchange energies, and M the degree of polymerization.

$$R_f = \left(|A_1|Mk\frac{L}{u}\right)^{-1}\ln\left[1 + \left(|A_1|Mk\frac{L}{u}\right)\exp[|A_1|M(X_{1m(0)} - X_{1mc})](1 - R_f)\right] \qquad [6]$$

where k is a rate constant indicative of the change in mobile phase composition. In the above case the rate law is considered to be independent of the composition of the better polymer solvent. $X_{1m(0)}$ is the composition of the mobile phase in the reservoir of

Figure 11 Illustration of the change in configuration of a polymer which occurs in the transition region (i.e., the region where polymers are not completely adsorbed or desorbed). The range of the transition region increases as a polymer's molecular weight decreases (9).

the development tank, L the distance advanced by the solvent front, and u the average rate of migration of the solvent front.

Equations [5] and [6] describe analogous processes. The main difference is that in LC separations (Eq. [5]) fractionation occurs by going from a poor to a good mobile phase, while in TLC (Eq. [6]) fractionation occurs by going from a good to a poor mobile phase as the chromatogram develops. In both cases the relationship between the mobile phase composition and the critical composition controls the elution of a polymer.

Both Eqs. [5] and [6] can be used to predict qualitative chromatographic trends. The closest correlation between theoretical and experimental values (of t_r and R_f) is for the higher molecular weight polymers (9). Indeed, the concept of a critical concentration is most accurate for the higher molecular weight polymers. For lower molecular weight polymers there can be an appreciable transition range in which the polymer elution behavior varies over a wide range of solvent compositions (Figure 11). In these cases one can envision more gradual transitions from the desorbed to the adsorbed state (Figure 11). Prediction of retention times from Eq. [5] tends to be more accurate than the prediction of R_f values from Eq. [2] (6). The major reason for this is believed to be that the solvent gradient is known and exactly controlled in LC but not in TLC. In addition, TLC is subject to changing phase ratios, wetting phenomena, etc. Despite these deficiences, the major theoretical trends are consistent with experimental observation.

Equation [6] can be solved directly (10); however, it is somewhat cumbersome and difficult to visualize in terms of experimental

results. Fortunately, one can simplify the expression to a more understandable form when analyzing certain special cases. For all but lower molecular weight polymers (where $X_{1m(0)} - X_{1mc} > 0$),

$$\left(|A_1|M \frac{L}{u}\right) \exp[|A_1|M(X_{1m(0)} - X_{1mc})](1 - R_f) >> 1$$

and Eq. [6] simplifies to

$$R_f = \frac{u(X_{1m(0)} - X_{1mc})}{kL} + \frac{u}{|A_1|MkL} \ln\left(\frac{|A_1|MkL}{u}\right) + \frac{u\ln(1 - R_f)}{|A_1|MkL} \qquad [7]$$

For moderate to high molecular weight polymers the second and third terms of Eq. [7] become increasingly small relative to the first term. Therefore, for "high polymers":

$$R_f = \frac{u(X_{1mc(0)} - X_{1mc})}{kL} \qquad [8]$$

When $X_{1m} - X_{1mc}$ is greater than zero (i.e., the mobile phase in the developing chamber is a good solvent), polymer migration will occur. The extent of this migration can be calculated provided the gradient (k) and the critical mobile phase composition (X_{1mc}) are known. X_{1mc} increases monotonically with M and asymptotically approaches a limiting value X^{∞}_{1mc} as $M \rightarrow \infty$. This indicates that chromatographic resolution should decrease as the molecular weight increases.

When $X_{1m(0)} - X_{1mc}$ is less than zero (i.e., the mobile phase in the developing chamber is a poor solvent mixture) then

$$\left(|A_1|Mk \frac{L}{u}\right) \exp[|A_1|M(X_{1m(0)} - X_{1mc})](1 - R_f) \rightarrow 0$$

for higher molecular weight polymers and $R_f = 0$.

Table 7 is a comparison of theoretically calculated R_f and t_r values with experimentally determined values for different molecular weight polystyrene standards. This is one of the few times a thin-layer chromatographic theory has been tested this rigorously and performed this well.

Table 7 Retention Times and R_f Values of Polystyrene Calculated from Eqs. [5] and [6] and Measured Experimentally

	t_r (min)		R_f	
M	Theoretical[a,b]	Experimental[c]	Theoretical[a,b]	Experimental[d]
22.38	3.1	3.2	0.965	0.715
166.67	6.8	8.3	0.742	0.568
333.33	8.5	8.8	0.597	0.507
1,047.60	10.4	10.1	0.418	0.429
3,714.30	11.8	11.6	0.300	0.358
35,238.00	13.1	12.7	0.209	0.244
95,238.00	13.3	13.3	0.185	0.185
	13.7	—	—	—

[a]Calculated using the following values of reduced energy parameters: $X_{12} = 0$, $X_{13} = 0$, $X_{23} = 0.5$, $\beta\Delta\varepsilon(1,3,5) = -0.102$ (see Ref. 10).

[b]The temporal variation of the composition was chosen to be $X_{(m)(t)} = 0.4163 + 0.031t$. The composition $X_m(\sigma) = 0.829$ corresponds to that obtained from $X_{(m)(t)} = 0.416 + 0.031t$ for the longest retention observed in the experiment outlined in footnote c. The value of K was chosen to match the experimental R_f value for the highest MW polymer.

[c]t_r values obtained from LC fractionation of polymer using a linear gradient beginning at t = 0 with a mobile phase composition of 53:47 (v:v) methylene chloride/methanol and increasing to 100 $MeCl_2$ in 18.8 min using ODS-3 column (length, 25 cm); flow, 1 ml/min.

[d]R_f values obtained from TLC fractionation of polymer using 79:21 (v:v) $MeCl_2$/MeOH.

It is apparent that the theoretical description is consistent with experimental observation. The results of this union are threefold:

1. The fractionation mechanism is largely understood.
2. A useful polymer fractionation technique can now be utilized and improved systematically.
3. Various forms of TLC and LC can now be used as a potentially powerful tool in the theoretical and physicochemical study of polymers and their behavior in various environments.

ACKNOWLEDGMENT

Support of this work by the Department of Energy, Office of Basic Energy Research (DE-AS0584ER13159), is gratefully acknowledged.

REFERENCES

1. D. W. Armstrong and R. E. Boehm. J. Chromatogr. Sci. *22*, 378 (1984).
2. W. J. Langford and D. J. Vaughn. J. Chromatogr. *2*, 564 (1959).
3. H. L. Spell and R. D. Eddy. Am. Chem. Soc. Chicago Meeting *24*, 2, 267 (1964).
4. H. Inagaki. In: *Advances in Polymer Science*, vol. 24, p. 189, Springer-Verlag, New York, 1977.
5. B. G. Belenkii and E. S. Gankina. J. Chromatogr. *141*, 13 (1977).
6. B. G. Belenkii. Pure Appl. Chem. *51*, 1519 (1979).
7. B. G. Belenkii and L. Z. Vilenchick. In: *J. Chromatography Library*, Vol. 25, p. 361, Elsevier, Amsterdam, 1983.
8. D. W. Armstrong and K. H. Bui. Anal. Chem. *54*, 706 (1982).
9. D. W. Armstrong, K. H. Bui, and R. E. Boehm. J. Liq. Chromatogr. *6*, 1 (1983).
10. R. E. Boehm, D. W. Martire, D. W. Armstrong, and K. H. Bui. Macromolecules *16*, 446 (1983).

11. R. E. Boehm, D. E. Martire, D. W. Armstrong, and K. H. Bui. Macromolecules *17*, 400 (1984).

12. S. Simunic and S. Turina. Kem. Ind. *28*, 413 (1979).

13. T. Tanaka, N. Donkai, and H. Inagaki. Macromolecules *13*, 1021 (1980).

14. G. Glöckner and R. Meissner. Acta Polym. *31*, 191 (1980).

15. T. Kotaka and J. L. White. Macromolecules *7*, 106 (1974).

16. K. H. Bui. Reverse-Phase Liquid Chromatographic Fractionation of Polymers, Ph.D. Dissertation, Georgetown University, Washington, D.C., 1983.

17. H. Inagaki, H. Matsuda, and F. Kamiyama. Macromolecules *1*, 520 (1968).

18. B. G. Belenkii and E. S. Gankina. Dokl. Akad. Nauk SSSR *186*, 857 (1969).

19. B. G. Belenkii and E. S. Gankina. J. Chromatogr. *53*, 3 (1970).

20. K. Kamide, S. Manabe, and E. Osafune. Makromol. Chem. *168*, 173 (1973).

21. N. Tagata dn T. Homa. J. Chem. Soc. Jpn. (Nippon Kagaku Zasshi) 1330 (1972).

22. S. Mori and T. Takenchi. Kobunshi Kagaku *29*, 383 (1972).

23. K. Kamide, T. Okada, T. Terahawa, and K. Kaneko. Polym. J. *10*, 547 (1978).

24. S. Teramachi and H. Esaki. Polym. J. *7*, 593 (1975).

25. T. I. Min. Pollimo *2*, 146 (1978).

26. E. S. Gankina, S. I. Ganicheva, V. M. Belaiev, J. J. Kever, and B. G. Belenkii. J. Chromatogr. *208*, 381 (1981).

27. J. Janca, J. Kroupova, S. Pokorny, J. Stehlicek, J. Kalal, E. S. Gankina, E. E. Kever, and B. G. Belenkii. J. Chromatogr. *171*, 81 (1979).

28. M. L. Langhorst. Dow Chemical Co., Midland, Mich., 1985 submitted.

29. G. Glöckner. Pure Appl. Chem. *55*, 1553 (1983); G. Glöckner, J. H. M. van den Berg, N. L. J. Meijerink,

T. G. Scholte, and R. Koningsveld. Macromolecules *17*, 962 (1984).

30. F. Kamiyama, H. Inagaki, and T. Kotaka. Polym. J. *3*, 470 (1972).

31. K. Kotaka, T. Uda, T. Tanaka, and H. Inagaki. Makromol. Chem. *176*, 1273 (1975).

32. H. Inagaki, T. Kotaka, and T.-I. Min. Pure Appl. Chem. *46*, 61 (1976).

33. N. Donkai, T. Miyamoto, and H. Inagaki. Polym. J. *7*, 577 (1975).

34. B. G. Belenkii and E. S. Gankina. J. Chromatogr. *3*, 141 (1977).

35. F. Horii, J. Ikada, and I. Sakurada. J. Polym. Sci. Polym. Chem. Ed. *13*, 755 (1975).

36. T. Taga and H. Inagaki. Angew. Makromol. Chem. *33*, 129 (1973).

37. F. Kamiyama, H. Matsuda, and H. Inagaki. Makromol. Chem. *125*, 286 (1969).

38. B. G. Belenkii, E. S. Gankina, P. P. Nefedov, M. A. Kuznetsova, and M. D. Valchikhina. J. Chromatogr. *77*, 209 (1973).

39. T. Miyamoto, N. Donkai, N. Takimi, and H. Inagaki. Annu. Rep. Res. Inst. Chem. Fibers. Jpn. *30*, 27 (1973).

40. H. Inagaki, T. Miyamoto, and F. Kamiyama. Polym. Lett. *7*, 329 (1969).

41. T. Miyamoto and H. Inagaki. Macromolecules *2*, 554 (1969).

42. T. Miyamoto and H. Inagaki. Polym. J., I, 46 (1970).

43. T. Miyamoto, S. Tomoshige, and H. Inagaki. Makromol. Chem. *176*, 3035 (1975).

44. H. Inagaki and F. Kamiyama. Macromolecules *6*, 564 (1974).

45. Y. Miyamoto, S. Tomoshige, and H. Inagaki. Polym. J. *6*, 564 (1974).

46. N. Donkai, N. Murayama, T. Miyamoto, and H. Inagaki. Makromol. Chem. *175*, 187 (1974).

47. H. Mazur and I. Lewandowski. Metody Badania Wyrobow Tworzyw Szlucznych. 3 (1977); Chem. Abstr. *93*, 72536X (1980).
48. T. Moldovai and P. Moldoveanu. Ind. Usoara *12*, 447 (1978); Chem. Abstr. *91*, 21654K (1979).
49. A. H. Lawrence and D. Ducharme. J. Chromatogr. *194*, 434 (1980).
50. H. Halpaap and K. Klatyk. J. Chromatogr. *32*, 80 (1968).
51. H. Determann. *Gel Chromatography*, 2d ed., Chap. 2, Springer-Verlag, Berlin, 1969.
52. E. P. Otocka, M. Y. Hellman, and P. M. Muglia. Macromolecules *5*, 227 (1972).
53. N. Donkai and H. Inagaki. J. Chromatogr. *71*, 473 (1972).
54. E. P. Otocka and M. Y. Hellman. Macromolecules *3*, 362 (1970).
55. F. Kamiyama, H. Matsuda, and H. Inagaki. Polym. J. *1*, 518 (1970).
56. E. P. Otocka. Macromolecules *3*, 691 (1970).
57. H. Inagaki, F. Kamiyama, and T. Yagi. Macromolecules *4*, 133 (1971).
58. E. P. Otocka, P. M. Muiglia, and H. L. Frisch. Macromolecules *4*, 512 (1971).
59. T. Taga and H. Inagaki. Angew. Makromol. Chem. *33*, 129 (1973).
60. E. P. Otocka. In: *Quantitative Thin Layer Chromatography*, J. C. Touchstone (ed.), p. 45, Wiley, New York, 1973.
61. B. G. Belenkii, E S. Gankina, M. B. Tennikov, and L. Z. Vilenchik. J. Chromatogr. *147*, 99 (1978).
62. F. Kamiyama and H. Inagaki. Bull. Inst. Chem. Res. Kyoto Univ. *62*, 393 (1974).
63. D. W. Armstrong, K. H. Bui, and R. E. Boehm. J. Liq. Chromatogr. *6*, 1 (1983).
64. H. Inagaki. In: *Fractionation of Synthetic Polymers*, H. L. Tung (ed.), chap. 3, Marcel Dekker, New York, 1977.

65. D. W. Armstrong and K. H. Bui. Unpublished results, 1983.

66. K. H. Bui and D. W. Armstrong. J. Liq. Chromatogr. 7, 29 (1984).

67. D. W. Armstrong, J. N. Marx. D. Kyle, and A. Alak. J. Chem. Educ. *62*, 705 (1985).

68. K. H. Bui and D. W. Armstrong. J. Liq. Chromatogr. 7, 45 (1984).

69. D. W. Armstrong and P. Dinyarian. Unpublished results, 1984.

70. E. A. DiMarzio and R. S. Rubin. J. Chem. Phys. 55, 4318 (1971).

71. T. L. Hill. *An Introduction to Statistical Thermodynamics*, chaps. 14, 19, 21, Addison-Wesley, Reading, Mass., 1960.

72. P. Flory. *Principles of Polymer Chemistry*, Cornell University Press, Ithaca, N.Y., 1971.

Index

B

F

G

M

N

O

T

U

V

W

X